A SEARCH FOR COMMON GROUND

A SEARCH FOR COMMON GROUND

Edited by Peter Gould and Gunnar Olsson

Pion Limited, 207 Brondesbury Park, London NW2 5JN

© 1982 Pion Limited

ISBN 0 85086 093 8

Printed in Great Britain by Page Bros (Norwich) Limited

Acknowledgements

This volume started with a small conference at 'The Point That Divides The Wind', a promontory at Bellagio on Lake Como, once the site of a villa belonging to Pliny the Younger, a place of which he once wrote:

> "One is set high on a cliff ... and overlooks the lake ... Supported by rock, as if by the stilt-like shoes of the actors in tragedy, I call it Tragedia. It enjoys a broad view of the lake which the ridge ... divides in two ... From its spacious terrace, the descent to the lake is gentle."[1]

Today it is the site of the Villa Serbelloni, owned by the Rockefeller Foundation, and used as a Conference Center to further scholarship and thought. From the spacious terrace, the descent to the lake is still gentle.

We would like to acknowledge, first and foremost, the sympathetic support of the Rockefeller Foundation, the program committee, and the program coordinator Susan Garfield in the New York office. But, and more directly, we would also like to acknowledge with gratitude the presence of Roberto Celli, and all the staff of the Villa Serbelloni, for the many daily tasks of gentle grace that they performed for us. We wish only that we could think with the same care that they display in their daily rounds, and stand open to new thoughts as they stand open constantly to the needs of their guests.

E quindi uscimmo a riveder le stelle.

We would also like to thank Ronald Abler, of the Department of Geography, The Pennsylvania State University, for supporting all the efforts of getting a manuscript together; the retypings, copying, supplies, postage, and so on, elements in a backcloth that too frequently are taken for granted—unless they are not there.

Joan Binkley typed the manuscript in all its many stages and revisions, and we thank her for her patience and skill throughout the long process.

Finally, we would like to thank Robert Graves for permission to reprint his poem *The Cool Web*, taken from *Robert Graves: Selected by Himself.*

Peter Gould
Gunnar Olsson

---

[1] Pliny the Younger, *Letters* (IX.7).

The contributors

Dietrich Bartels — *Geographisches Institut, Institut für Regional-forschung der Universität Kiel, Olshausenstrasse 40–60, 2300 Kiel 1, West Germany*

Kathleen Christensen — *Environmental Psychology Program, The Graduate School and University Center of the City University of New York, 33 West 42 Street, New York, N.Y.10036, USA*

Helen Couclelis — *vas. Georgiou B' 15, Athens 138, Greece*

Reginald Golledge — *Department of Geography, University of California, Santa Barbara, California 93106, USA*

Peter Gould — *Department of Geography, Penn State University, University Park, Pennsylvania 16802, USA*

Derek Gregory — *Department of Geography, Cambridge University, Downing Place, Cambridge CB2 3EN, England*

Bernard Marchand — *75 rue Quincampoix, Paris 75003, France*

Gunnar Olsson — *Thunbergsvägen 22, S-75238 Uppsala, Sweden*

Anne Osterrieth — *Université Catholique de Louvain, Unité de Géographie Régionale et de Télédétection, Mercator-Place Louis Pasteur 3, B-1348 Louvain-La-Neuve, Belgium*

Allen Pred — *Department of Geography, University of California, Berkeley, California 94720, USA*

Allen Scott — *Department of Geography, University of California, Los Angeles, California 90024, USA*

Erik Wallin — *Spårsnövägen 14, 22252 Lund, Sweden*

Contents

Prologue: A search for common ground

Peter Gould

That geography is an eclectic discipline is well-known. That from the breadth of its inquiries spring its strengths and its weaknesses is equally acknowledged. The wide range of its concern gives it the potential for generating deep and important insights of considerable intellectual strength. In these days of increasing specialization, and what we might call 'partitional thinking', geography's tradition of eclectic and integrative inquiry represents a perspective of incalculable value both for the relatively sheltered world of the university, and for the larger, open society in which it is embedded. But the very source of geography's strength is also the origin of its weaknesses. If an older tradition of exploration is reinterpreted, then today the map of geography itself often appears fragmented and torn. Sometimes it looks as though the explorers have moved off in all directions, with little sense of how their individual findings fit into a larger concern for knowing. The research fragments too frequently lie scattered over the landscape of inquiry, like bits of a jigsaw puzzle, without connection, and so, by definition, without structure and pattern.

Part of the problem of fragmentation stems from the sheer explosive exuberance of the past quarter of a century. Despite local setbacks, in many countries there have never been so many students of geography; new journals seem to appear each year; books of worth, thoughtful reflection, and insight are published with regularity; and more geographers are engaged with practical problems of an urban, regional, national, and international nature than ever before. A search for deeper descriptive precision has been followed by an equal concern for the social context and implications of geographic research, and these broader questions have been translated into more pointed ones of an ethical and moral nature. It has been an exciting time to be a geographer, and to think and inquire about the world through one of the many, and intellectually provocative, lenses that shape the contemporary geographer's seeing.

At the same time, the explosive nature of geographic inquiry seems to have left too little time for critical reflection. For many it was enough to explore, and never mind which direction was chosen, or whether the messages carried in the forked sticks of the journals from one enthusiastic explorer to another got through. In any case, some of the explorations needed new languages to describe what was found, and even if the messages got through physically as printed pages, they often said little to those working in other regions. In the swamps, mountains, and deserts of the geographic landscape there seemed to be few philosophical trees casting their shade for explorers who wished to gather their strength, reflect upon what they were doing, or question the purpose of their last journey—or the next to come. Despite some possible assertions to the contrary, the tradition of philosophical reflection has never been strong—even in the 19th Century—and until quite recently much that has passed for it has lacked both insight and depth.

Along with almost every other area of human inquiry, geography has paid a severe price for failing to maintain a close association with an older tradition of philosophical reflection and discourse, an area of thinking where standards have often been high during the past 2500 years. The price has been paid in the coin of reflection, in the failure to maintain a tradition of larger philosophical concern to which we might repair from time to time to think, deeply, about what we are doing, and in the apparent inability to get 'on the outside looking in', and so see our professional selves and work in a larger, and intellectually enriching, context.

The conceptual, methodological, ideological, and ethical perspectives of the past twenty-five years have often emerged in critical opposition to the current ways of seeing and, without the guidance and tradition of explicating critique to clarify and widen the debate, they have also become yet more temporary dogmas. Labels such as *scientist* or *humanist* tended to caricature emerging positions, giving false comfort to small coteries whose introverted rise made wider, and more thoughtful communication even less likely. Explorers who plant banners on small foothills, and then proceed to thumb their noses at others across the valley, are unlikely to see the towering peaks of true understanding lying behind them in the mists and clouds, let alone realize that their scaling may require joint expeditions.

It was with such a desire to open clarifying discourse between geographers of different persuasions, that the authors in this volume (and one or two others) met in common concern. Without exception, geographic discussions were informed by a highly varied and often deep interest in contemporary philosophical writing. Not that discussions were intensely philosophical, as they might have been for professional philosophers, but there was a constant awareness of the way in which contemporary philosophical themes and writing underpinned and informed the geographic discourse. Many also became aware of the way the discussions, often punctuated by questions that were rhetorical, were themselves a reflection of the particular historical time in which they were being generated. Inevitably, we are part and parcel—even prisoners?—of our times; yet we also have the capacity to reflect upon this very fact. In such moments of awareness, moments which sometimes emerge in the midst of disagreement, the mist behind us thins slightly, and we get a glimpse of the high and steep slopes ahead.

When people are prepared to stand open to new possibilities, to new ways of seeing, we may hope for a contraction in the intellectual space in which they reside. We even made some attempt to demonstrate such a drawing together, asking (on a totally confidential basis) if each person would scale all the others in terms of 'professional intellectual affinity' to himself or herself well before, and immediately after, the five days of discussions. We had the idea of constructing somewhat oversimplified 'before' and 'after' maps, to see, in a literal graphic sense, how positions had changed and if the overall space had contracted[1].

Our experiment failed, in the sense that it was never completed. Nearly everyone took part in the 'before' scaling, but most were so uncomfortable with the task of repeating it after the five days of discussions that the experiment could not be finished. From the few, second round returns we received, it seems almost certain that there was an overall contraction—people felt intellectually 'closer'—but the scatter of scalings only confirmed the general impression, held by nearly everyone, that the chance to meet and talk, if not producing agreement, at least produced greater trust, awareness, respect, and even understanding for other points of view. One person likened the conference to an archipelago: the effects of our discussions were not to be measured by great forces of continental drift that drove the individual islands into a homogeneous landmass, but by the throwing and retrieving of gossamer filaments that allowed the construction of long suspension bridges. With such structural links in place, ideas can now travel more easily.

Structuring a book—traditionally a linear narrative—from the diverse interests and viewpoints represented by the twelve people who contributed to this volume has been no easy task. Not because the pieces represent disconnected fragments, but because so many themes and, therefore, connections in common generate so many possibilities. We could even think formally of a well-defined set of authors (A), and another well-defined set of themes (T), with a relation $\lambda \subseteq A \otimes T$ between them that could be conceived as a geometric space in which we, the editors, have the job of transmitting you, the reader, as traffic on a backcloth[2]. Given enough time, and many rounds of clarifying definitions, we could actually state such a relation, and use the resulting multidimensional geometry as a guide. But this would have delayed publication for a year, so we used our own judgment. What we do know, intuitively, is that the geometric space is not the conventional one of linear narrative—the individual essays do

[1] Quite apart from anything else, such a research task raises formidable conceptual and methodological problems. If spaces are created by multidimensional scaling, how *exactly* does one measure an overall, or even differential, dilation or contraction in the space? I commend this problem of map construction and comparison to those who enjoy wrestling with such things. I do not know how to solve it in a completely satisfactory way.

[2] Such an experiment was actually tried out in a somewhat different context, using the sets of authors and references in S Gale and G Olsson, *Philosophy in Geography* (1979 Dordrecht; D Reidel); see P Gould "The Structure of a Discourse Space: Some Pen-on-Paper Reflections on *Philosophy in Geography*", *mss.* privately circulated to the twenty authors.

not connect into a long chain—but something closer to what an algebraic topologist would call a *star*, with each essay connected to the others by themes that indicate varying degrees of shared concern. On such a complex structure, you, the reader, will find yourself returning to some basic themes again and again. Not, we hasten to add, in any repetitive sense—the diversity of thinking guarantees that there is little, if any repetition—but in the sense that each author will lead you to themes that may have appeared before, but will then discuss them from a quite different perspective.

It is hardly surprising that such themes as the individual and society, languages, dialectics, structures, contemporary philosophical thought, and the recent history of geography itself should emerge so frequently. They are part of a wider concern in our modern world, and we suspect that they will still be around in the centuries to come. What is surprising is the way they appear so often in the essays; sometimes fleetingly and tangentially, sometimes pointedly and directly, but always demonstrating to the people who thought they were on different islands in the archipelago that the strands of shared concern could be used to construct stronger bridges.

We start with two, very different views of recent developments in geography. Golledge is inherently optimistic, though not uncritical of some of the wider swings of the pendulum since the early 1960s. There is an acute realization that we could not be where we are today without having passed through some of the shifts in thought over the past two decades. The lasting contributions of geography in the United States during the 1960s include a much deeper and continuing concern for logical and methodological rigor, and a desire to create genuinely shared knowledge that *builds* upon secure foundations. Although sharing such values, the view from West Germany by Bartels is bleak. It seems that the 'competent' people, those with something to contribute in the wider public sphere, often leave the universities, while those who stay behind drift from one justification to another, or simply do their 'thing' without much sense of any common endeavor with their peers. There is an antiscientific trend that reflects both developments in the discipline, as well as larger movements within the society itself.

In characterizing trends in geography over the past twenty years, we use words that appear to have very different meanings. Many come from philosophical discourse, where their meanings may have been quite clear originally, but in transferring them to geographic discussion we have sometimes confused terms and meanings to the point that we often talk past one another. To clarify the meanings requires that we go back and examine carefully the original intentions of those who first coined them. This is precisely the task Christensen has set herself, distinguishing between *critique* and *criticism*, and opting for the former approach to push us back to the meanings of words used so frequently, and inaccurately, in the humanist-scientist debate. Both sides will benefit from this essay, and in the future perhaps we shall see a greater propensity to use the precise terms of philosophy with greater care, *shared* meaning, and enhancement of understanding.

Meaning is crucial: and a very different perspective comes from Osterrieth with her concern for lived space (*espace vécu*), particularly as it takes on meaning for those who live in large cities. We do not simply live *in* space, but interpret place, experience it, put value upon it, and give it symbolic significance. Our concern, at this highly individual level, is partly diagnostic: to assess the quality of lives lived by real people, and to evaluate the normative values that allow us to prescribe changes that make for a more decent and humane life.

Whenever we see two groups of people in confrontation over an either-or situation, there is always the temptation as an outsider to ask whether one really has to choose between the alternatives that appear so diametrically opposed. It is just this question that I would pose to those so firmly entrenched either in the scientific or in the humanistic traditions. Not only is science a thoroughly human enterprise, but it requires that most human of all characteristics—interpretation and the impression of meaning upon the things that we observe or create. Often we are engaged in a search for structure, and may need languages appropriate to the task: to operationalize our intuitive notions of structure that permeate much of our discourse, and to distinguish between structures that support and the things that are supported.

A concern for structure also underlies the essay of Couclelis, and is reflected both in her broad concern to find structure and coherence in the field of geography, as well as in her more sharply defined goal to explicate structural questions by using urban models as a particular example. Moving up and down a hierarchy of relations, between the most concrete and the most abstract levels, she invites us to consider a succession of perspectives on human inquiry, and in the course of her discussion shows how the addition or subtraction of information precipitates change, so altering our views of space, and perhaps of time itself. Though much of her thought, particularly as it is expressed technically, may induce some to label it as purely in the scientific tradition, her final plea to realize the fundamental human role in structuring reality asks us to eschew the old partitions that trap thinking and limit its scope.

Whatever we write or think, we do it in a social matrix that characterizes a certain historical era. Even as we try to step back, and attempt to get 'on the outside looking in', we realize that we have been born and have grown up in a time when precisely such an emancipatory activity is possible. The recursive step is always one step ahead of us. In one of the most provocative essays of the book (an essay for which we all agreed to set aside a whole day, producing discussion that informed and influenced all subsequent discourse), Scott invites us to think about *why* we think so intensely today about the spatial aspects of our society. He argues that knowledge and science are historically determinate, and that in the areas of urban and regional discourse particularly we inevitably focus upon the problems that reflect social and property relations of late capitalist society. It is an essay written in the form of an invitation: an invitation to the reader engaged in geographic inquiry to be self-conscious about his or her position and role in the society under investigation, and to be critically vigilant as one attempts to connect thinking and insight to the social and political matrix in which such investigation takes place.

The relation of an individual to the larger society is also the subject of the essay by Pred, but it is stated from a very different perspective as he places a particular tradition of geographic research in a wider, and shared concern for social theory. A major goal of time-geography is to detail the time-space constraints on the individuals that make up a particular society, with all its physical 'plant', institutions, and projects. Yet at the same time that society shapes and constrains an individual, so the sum of individuals makes up society—in an on-going dialectical process that continues minute by minute and lifetime by lifetime. And language itself is dependent upon social reproduction—the constant reproduction of society by individuals—even as social individuals require, and in a sense are defined by language.

The concern to link contemporary human geography with modern social theory is exemplified by the essay of Gregory, a statement remarkable for its equally broad grasp of geographic foundations and nuances of current social debate. Not only does he illuminate some of the old geographic controversies and their harshly drawn dichotomies, but he reminds us of the degree to which early members of the locational school were intimately aware of the social implications of their theoretical work. Distant peaks of research are not only hidden from explorers by mists, but also by the thick undergrowth of misunderstood arguments. Gregory's essay represents an attempt to clear some of the old and new thickets away, so demonstrating how modern human geography can provide an expanded and enriching perspective to inform current thought in social theory.

Those who have followed the searching essays of Olsson over the past ten years are more than simply aware of his concern for language[3]. With James Joyce as his companion, he has asked us by example to *think* about the very medium in which we think, a medium without which we cannot think, yet a medium which shapes our thinking—even as we think we are thinking about it. As always his essay is a personal statement, in a highly personal style, and if we are prepared to stand open, and read it with the same care as it was written, we can feel with him the joy of knowing that we do not know, and the agony of knowing that we do. As you read his

[3] A number of these have been brought together in the coming-from-the-other-way collection that forms the second part of G Olsson, *Birds in Egg/Eggs In Bird* (1980 London; Pion).

essay, gather weigh slowly and stop frequently on the way; it is the only way to weigh.

Language is not simply verbal, and dialectical relations expressed in many languages appear in many guises in geographic space—provided we have eyes to see, and a framework that gives meaning to what we see. The essay by Marchand takes us from general considerations and abstract concern to a direct and concrete example of dialectical thinking in an urban area—Los Angeles, California. There he unfolds the relations between the values of use, exchange, sign, and symbol, showing how they turn in upon themselves, and help us to form associations that bring out the deeper structure of the urban landscape. It is a highly original perspective on the urban scene—perhaps a viewpoint that could only come from one who knew his Roland Barthes, and was both outside and inside the California society at the same time.

The final essay by Wallin is also embedded in questions of language, but in a way that may not be immediately apparent to the casual reader of English. His writing and thinking were originally in Swedish, a language that initially appears lean and spare compared with English, with its rich roots both in the Germanic and the Latin streams, yet a language in which Wallin demonstrates an almost Heideggerian etymological skill. The plays on words, the meanings at once so seemingly ambiguous, yet so precise, are used to illuminate a concern for what he has termed *passages*—the transitions in human lives, where one thing turns into another.

Our hope now is that these essays will encourage others to take up some of the broad questions they pose; either as individuals sitting quietly in the evening, reflecting upon their past, current, and future research and teaching, or as groups, perhaps in graduate seminars, who may wish to use the essays as points of departure for their own discussions and thinking. In the course of undergraduate and graduate work, it seems to us that there are too few opportunities to lay out and discuss questions that are not only of critical importance to professional geographers, but often to the wider range of people who deal daily with the human condition—which means, virtually by definition, the larger society. Above all, we hope the concern shown here will encourage geographers to read more widely both in contemporary and in older traditions of philosophy. There is no curriculum here: people must find their own ways, their own paths through the forest—although good teachers, as always, can be an enormous help. We commend a greater awareness of the tradition of reflective thinking at its best. It is possible that if such a tradition had been stronger in the past, then present issues would not be seen in such hard, confrontational terms.

Optimism and pessimism in contemporary geography

Fundamental conflicts and the search for geographical knowledge

Reginald Golledge

Confusion and uncertainty, or richness and diversity?
Among many geographers, there is a deeply felt concern for the current state and future of their discipline. Such concern, while being common to members of many disciplines today, is very apparent in geography as societal priorities are manifest through the closure of academic departments at major educational institutions, a slackening student concern, and the erosion of the geographical basis of local, regional, and global knowledge in many school systems. Teachers and researchers alike perceive a lack of thrust in the discipline, with no clear path to follow, multiple directions to choose from, and few commonly accepted principles to guide them in the search for understanding and knowledge. The result is a discipline exhibiting a degree of confusion in its daily practice, characterized by fragmentation of thought, controversy over varying beliefs and ideals, and uncertainty with respect to its role in regard to environment, individual, society, system, institution, ideology, theory, and practice. Of course, an optimist would point out that this state of affairs is indicative of a vibrant discipline that is not oriented to the *status quo*, but which constantly evaluates itself in an attempt to improve its knowledge base and its consequent level of understanding of the nature of things. In the view of this latter group, geography today is exhibiting a richness and diversity beyond any previous period of its existence. This essay is devoted to explaining why I adhere to this latter optimistic position. To do so, I shall comment on the particular views of knowledge and the intra-disciplinary conflict that underlie my beliefs.

The malaise said to exist in the discipline is, I suspect, felt more by our senior than our junior researchers. In particular, I suggest that it is the 'junior' senior researchers, those now approaching middle age who were part and parcel of the several revolutions of the 1950s and 1960s, who see the greatest amounts of confusion, and who suffer the greatest bouts of uncertainty. These are people who may have pushed as far as their current training and abilities can take them, who are dissatisfied with where they are today, but are uncertain as to what to do about it or where to go. Whether it be a 'passage' in the life of this group, a menopausal directive to reevaluate and assess, or a disenchantment arising from twenty years of work that has not produced expected ends, it is perhaps inevitable that some dissatisfaction should result. In contrast, many young researchers in the discipline are active, aggressive, full of ideas and ideals, very insightful, and *extremely* excited by the possibilities that exist within the discipline as it is today.

Regardless of the degree of optimism, there are a series of fundamental questions that plague each and every one of us. Are we in pursuit of something in order to exist? How can we reconcile the things we *actually* do to ensure a comfortable daily existence with ideals and visions of what geographers *can* or should do? Do any of us have a clearly defined set of goals, ideals, and aims that transcend the limited problem-solving that permeates our everyday lives and activities? Do we judge the success of our efforts by results that can be applied, evaluated, and measured in terms of their impact on humanity and its environments? Or do we really judge by the degree of comfort that our work allows us to extract from the system in which we live?

Unlike many other geographers who at this time are governed by despair, I am not. I see geography on the verge of making its greatest contributions to knowledge, on the verge of becoming an integral and important part of other labelled subsets of knowledge to a much greater extent than has ever been the case. I see other areas of knowledge turning to geography, not just for factual information, but for vitality, for a fresh perspective, for a way of generating thoughtful ideas, and for adding dimensions to their own specialized subareas that have heretofore been neglected.

However, when I look at some areas of the discipline, I too feel uncertainty and disquiet. I see soaring edifices based on the most flimsy foundations. I see technology and techniques rampant, often with inadequate thought to their relevance. I see a lack of concern for defining problems carefully, for clarity in defining terms, and for extracting meaning from the associations and relations that are hypothesized or are recovered as a result of research activities. I see small problems being worked on, many of them in very unimaginative modes, and even smaller answers being found to such small questions.

Much contemporary geographic thought is concerned with presenting us with evermore refined and evermore abstract models of humanity, society, the individual, and the many and varied systems in which we live. Such models are supposed to be not only more elegant from a technical point of view, but also more relevant in their representation of the phenomena that they purport to represent. Formalizing such models has given us many things to think about. For example, if we look closely at the models of the 1950s and 1960s it is almost inevitable that an attack should have been launched on the behavioral assumptions of even simple models. The physical assumptions were not attacked because they were all so obviously incorrect that no one bothered much about them. In the 1970s, the same models were reexamined from the viewpoint of the societal assumptions that were built into them, and the concepts, which were inherent both in their formalization and in their expected outcome, were critically assessed.

Each of the last few decades has seen the development of a major internal controversy in geography similar to those mentioned above—in the 1950s, a positivist scientific view was espoused that offered an alternative to the unique, holistic, and regional approach; in the 1960s, the quantitative and theoretical 'revolution' polarized the discipline; and now, in the 1970s, various humanist ideologies have been proposed as an alternative basis for the search for knowledge. But these latter movements, growing in part out of a concern for 'relevance' that emerged in many disciplines in the 1960s, carry with them a strong concern for 'realism' and for 'usefulness', which has forced the discipline to reevaluate what it means by 'advancing knowledge'. In particular, the emphasis on practicalities, an insistence on the need to comprehend societal and institutional constraints on individual and group spatial behavior, and a search for guidance in dialectical materialism, have added to the spread of uncertainty and disquiet among segments of the discipline. Although these changes have contributed even greater richness and more diversity to geographic thought, they have also inevitably raised questions of the possibility of excessive fragmentation, and the consequent loss of disciplinary identification and coherence.

In geography today, a major problem is that of detecting "signals in the noise" (Gould, 1979). To some extent geographers believe that noise is growing disproportionately to signals, and the consequent increased difficulty of signal detection causes frustration, argument, complaints about a lack of guidance and direction, and an incentive to search for 'common ground'. The increase in 'noise' is often attributed to excessive fragmentation of the discipline, to the point where it actually obstructs the passage of "fruitful, creative and original ideas, methodologies and perspectives" (Gould, 1980). It is true that excessive fragmentation may obstruct the passage of ideas, but it may also act as a filter ensuring that only the more durable and universally accepted ideas will diffuse widely. Most geographers admit that divisionalism and fractionalization are part of our discipline. But this is not necessarily 'bad'—for what is needed is not uniformity of philosophy and thought, but a "greater intellectual respect for, if not full agreement with, other points of view" (Gould, 1980).

Thus the fragmentation of knowledge and interest in the discipline of geography is certainly not an unusual event. By definition, as knowledge expands there is greater and greater difficulty in comprehending it in its entirety, and there is inevitably some increase in the tendency for specialization to take place. Unfortunately, specialization is often accompanied by a rather perverse view that a given area of specialization is the *only* area that is worth pursuing! Other areas, other interests, other approaches, and other philosophies are dismissed or downgraded because they are not central to the particular specialist's way of thought.

In geography, where a tradition of broadscale and holistic thinking dominated for many years, ever-increasing specialization has had both good and bad effects. It has helped refine some broad and disturbingly vague concepts that are central to our discipline, and it has shown that our peculiar mix of interests in the physical and the human makes us far more susceptible to schizophrenia than are other areas of formal inquiry. Our concern with external physical and social environments had led us to embrace science and scientism; our concern with sentient beings has led us to embrace humanist philosophies, in which beliefs and values and free will and society all interact and show us how difficult it is to adopt a rigorous scientific point of view. And now our concern with usefulness and relevance appears to be emphasizing that the ideological bases of research control our level of potential understanding.

But are we any less fragmented or any less philosophically diverse than we were twenty years ago, when the series of modern 'revolutions' shook the discipline and inevitably pushed it in the direction of its current uncertainty? Why is it that fifteen years ago many geographers thought they knew who they were and where they were going, whereas few today would admit to the same? Perhaps the answer lies in a clouding of the sense of 'belonging', and a loss of self and disciplinary identification.

A discipline that has no goals, no purpose, and no order is, by definition not a discipline. Rather it is replaced by the undirected wanderings of individual minds. In the 1960s, for better or for worse, positivist thought gave geographers a purpose and focus in a discipline that was fragmented to such an extent that it had little confidence in its ability to produce meaningful work. As a philosophy it encouraged logical thinking, objective measurement, and the search for generality: it also introduced a hope that geographic theory might exist. As such it imposed a firm structure where little structure existed; it brought 'respectability' with peer groups in other disciplines; it substituted discipline-wide purpose for individual flights of fancy; it suggested ways of organizing information, and subjecting such information to a variety of reasonings; it provided a *raison d'être* for continued purposeful investigation—namely, to achieve levels of explanation of spatially distributed phenomena that had not been formerly achieved.

The result was an explosion of knowledge; an explosion that helped lay bare the inadequacies of many ill-formed normative theories; an explosion that sent geographers deep into the knowledge of other disciplines to aid their comprehension of spatial phenomena; an explosion that revealed a multitude of *other* philosophies with which geographers could identify, and by whose principles they could pursue their own roads to achieving understanding. And most of all, it rescued geographers from decades of constructing storehouses of facts, which, like grain, tend to become mouldy and rot over time.

In short, the 'scientific' approach, closely allied with positivism as a philosophy, attempted to make one think rigorously, although in specified channels rather than in a wide screen. It also provided a philosophical aim for much of the research associated with discovering distributions and identifying their characteristics; it searched for meaning by concentrating on discovering connections amongst facts, rather than fact-finding itself; it clarified the concept of inference for geographers; it introduced ideas of reliability and objectivity to the discipline; and it tried to provide a mechanism for distinguishing significant from meaningless material.

It would be unjust to claim that positivism and the use of scientific method achieved all that it aimed to do. In fact, it might be said to have been the *mechanism* that inhibited many of the 'dreamers' of the discipline. In many cases, the use of the scientific method did nothing more than show that concepts were poorly defined, that much research effort was repetitive and unrewarding, and that many spatial problems, particularly those related to human spatial behavior, may not be *fully* examined using positivist principles.

The rigor and objectivity commonly associated with the scientific method is not the exclusive prerogative of the philosophy of positivism. The last decade has shown that it is possible to embrace various 'radical' philosophies while maintaining a desire for objectivity and communicability in the search for knowledge. Researchers can deeply concern themselves with societal and disciplinary reform, but adhere to the principles of scientific method in discovering areas of needed change, and in developing recommendations for change. This has produced a feeling that the discipline should not be dominated by one mode of thought, and that perhaps it is the existence of such 'divisions' in geography that has the potential to make it one of the more imaginative and exciting disciplines of the current decade.

The nature and acquisition of geographic knowledge

"Knowledge is an adaptive activity of an organism, it is a progressive adjustment of one part of the flux of process to other parts, and in the reaction, the part reacted to becomes environment in relation to the part reacting" (Lee, 1968, page 168).

Knowledge does not come in ready-made units which are impressed upon the mind as upon a 'tablula rasa'. Instead, it is created through the development of percepts, concepts, and language. Percepts become clearer and more precise as the involved concepts become clearer and more precise, and these become clearer and more precise in turn as they are satisfactorily translated into language. A language allows boundaries to be imposed on concepts via the media of definition. It also allows concepts to be passed on as knowledge, and allows percepts to evolve, as the critical components of the concepts are understood through the various cognitive processes. Knowledge, therefore, accumulates at the individual level, and by making percepts precise it allows an opportunity for previously unseen connections to be made. These connections can then be formalized and transmitted, through the media of language, to humanity in general. Recall that as adult humans we live in a world of ordinary perceptual objects, which we are able to perceive through our various senses and identify because we have learned the language which includes the relevant concept.

In other words, we receive a set of messages or perceive intuitive data; we identify selected characteristics of the messages or the data; we interpret the messages according to a language; and then comprehend the message involved in the data set. Such a sequence represents an 'acquisition' of knowledge.

"All knowledge is hypothetical in some sense or another. Exactly in what sense each kind of knowledge is hypothetical must be determined. Each kind will be found to be characterized by particular interplay between percepts and concepts" (Lee, 1973, page 17). There is, therefore, no single road to acquiring knowledge, to furthering understanding, and to achieving a more satisfying existence. To presume that there is either shows cupidity or ignorance. To suggest one way rather than another takes the courage of one's convictions, but to admit that one has been too singleminded in a quest for knowledge shows an increased level of understanding of the complexities of life and existence. To repudiate a period of learning because it was motivated by a philosophy which one might not now hold, or because it is based on a set of premises since proven false, understates the contribution of that period to the acquisition of knowledge.

The growth of knowledge is a continuing process which takes many strange directions and has many odd way stations. To argue that some of these directions are fruitless, and that some of the way stations are unnecessary, is to assume that the path of knowledge acquisition is clearly known and that experimentation is valueless. This is clearly not the case in geography.

There is, then, one thing of which we can all be sure—that there is no single way to obtain knowledge. Geographers have had their arguments on the relative worth of deduction, induction, the scientific method, logical positivism as a guiding force, the relevance of statistics and mathematics, the value of humanistic thought, and the value of radical ideologies: one could prepare a large list of the debates upon which geographers have focused their attention during the last two decades. But such a list is exciting! It reflects not only the rapid growth of the discipline as a whole, but it illustrates a depth of concern and interest that is a tribute to the discipline and its desire for expanded knowledge. After all, what is it that we are striving for? Is it not knowledge? If we were simply to understand *what* knowledge we have, and *how* each of us may contribute to expanding it, then perhaps we will have found the common ground we are searching for. The questions to be asked, therefore, are simple: What is knowledge? How can we contribute to knowledge? How can we communicate those contributions to others in such a manner that they can be appreciated and understood? And not the least important question:

are there any particular problems facing geographers which inhibit their acquisition of knowledge?

Just as it is appropriate that members of our discipline concern themselves with trying to upgrade the quality of society by immediate social action, so too is it proper to allow others to meditate on the meaning of spatial existence, or to try to organize their thoughts in a manner meaningful to themselves and others, or to describe according to clearly stated assumptions a normative world whose operational purposes are well defined. Without the first group, static unimaginative societies may evolve; without the second, the purpose of existing may never become clear; and without the third there would be fewer recognizable choices of alternative universes that may ultimately come into existence.

My point is simple, and it reflects my thoughts of almost a decade ago (Golledge, 1973). Whether knowledge is acquired (1) through deep personal involvement with society, its structure, its operation, and its problems, (2) by meditation and synthesis of our limited existing knowledge, or (3) by spinning webs of logical inference in an isolated ivory tower, it is, luckily for us, still very much a matter of individual choice. *And so it should remain*, for such freedom of choice provides the richness and diversity that distinguishes a progressive discipline from a collection of thematically arranged ideas. To exhort geographers to pursue *praxis* rather than theory is to deny a potential path to achieving knowledge. Although constant assessment of the state of knowledge and reevaluations of guiding philosophies are a necessary part of the accumulation of knowledge and understanding, let us not too readily embrace the principle of dogmatic rejection lest we replace one 'inadequate' dogma by another equally as poorly suited to the pursuit of knowledge and the well-being of man.

In many disiplines the continuing development of knowledge has had, as its principal concern, the binding together of hypotheses and laws into systems of theories. As Bergmann put it (1958, page 35):

"As a science develops, a store of laws and concepts accumulates together with an awareness of some connections, deductive among the laws, definitional among the concepts. At a certain point in the development it pays to arrange this material into theory."

When the social or behavioral scientist speaks of a 'basic orientation', or 'methodological approach', or a 'general theory' within which he is working, he generally has in mind a set of procedural rules that he claims to follow (Brown, 1963), and he usually recommends the set of rules for dealing with certain types of problems (for example, Berry et al, 1963; Ackerman, 1963; Golledge and Amedeo, 1968). The stuff from which human relations and social structures are made is not intuitively as evident as are the parameters and variables (such as length, time, and force) available to the physical scientist. Consequently, an effort must be made to distill or abstract such stuff from innumerable events. The selection of such events depends to a large extent on one's experiences, cultural back-

ground, and biases (Rapoport, 1959). Despite such limitations, attempts
are still made to select fundamental entities within the field of interest,
but the process of selection is frequently so involved, laborious, and
difficult that it constitutes the bulk of a research effort. *It is not
surprising, therefore, that many researchers hardly ever get around to
stating postulates, let alone formalizing a theory in the strict philosophical
sense of the word.* Much time is spent relating vocabularly terms to some
type of referrent, which must be abstracted from a rich variety of events,
generalizations, and relations in turn. By the time some of the referrents
have been abstracted and named, there is already a bulky system in
existence, which makes the task of seeking out generalizations, and
combining these into theories, an onerous one. In some disciplines such a
task is hardly ever begun. It seems to me to be very much a state of the
art of geographic research that most researchers are still searching for
referrents, and trying to distill from the immense amount of data in
human and physical environments some critical concepts, terms, and
events, rather than attempting to construct new theory.

A well-validated theory systematizes, or logically interconnects, isolated
knowledge through its system of hypotheses. It also tries to give an
explanation or account of these hypotheses, and serves as an integrating
mechanism for existing knowledge in its domain. Many social and
behavioral theories (not just those within geography) are embryonic in
nature and look more like loosely connected hypotheses, some of which
may be only partially tested. Many will not progress beyond this point.
Like the well-formed theories of physical science, however, such theories
aim at guiding the search for new knowledge, drawing attention to what
has been overlooked in the past, discovering past errors, ambiguities, and
gaps, and helping to ensure that the occurrence of these in the future is
minimized. It is not necessary to believe a theory to find it useful.
Despite recent pressures on many sides for academics to become more
interested in the 'empirical problems of the real world', there is no doubt in
my mind that meaningful research must continue to rely either on a sound
existing theoretical base (at least as sound as the discipline can provide) or on
an integrated and legitimate sequence of hypotheses which require testing in
well-defined experimental designs in order to produce meaningful results.

To support this position, I would like to discuss briefly some of the
ways that theories contribute to research and understanding. As Mario
Bunge has suggested (1967, pages 380–381):

"The childhood of every science is characterized by its concentration on
the search for singular data, classifications, relevant variables, and isolated
hypotheses establishing relationships among these variables, as well as
accounting for those data. As long as a science remains in this semi-
empirical stage, it lacks logical unity: a formula in one department is a
self-contained idea that cannot be logically related to formulas in other
departments of a science."

I have stressed that theories organize and systematize bodies of knowledge, and that they do this by carefully itemizing the relationships between bits of knowledge, whether these be statements of fact or lawlike statements. Once the basic theorized relationships are clarified, further knowledge accumulates by unpacking sets of *interrelationships* between the bits of knowledge and examining their meaning and use. If a theory about a given area of knowledge exists, it serves as a datum to which we refer new facts and new sets of relationships. Thus research findings can be examined for meaning and significance by relating them to theory.

An additional comment must be made concerning the interface between theories and empirical evidence. A fundamental dilemma arises when examining this interface:

"The use of theoretical terms in science gives rise to a perplexing problem: Why should science resort to the assumption of hypothetical entities when it is interested in establishing predictive and explanatory connections among observables?" (Hempel, 1965, page 179).

The solution to this dilemma lies in the fact that, in order to make a meaningful transition from observable data to prediction in terms of the observables, there needs to be some systematic connections effected by statements making reference to nonobservables. In physics, for example, connections are made between the volumes and weights of solids and liquids by referring to the hypothetical entity 'specific gravity'. Geographic researchers probably make at least as much use of such nonobservable relations, for in order to think about theoretical connections they must also make a detour through a domain of not directly observable things, events, and characteristics—all of which make it harder to specify the nature of connection between observables and prediction.

Hempel also warns us that a "deductive system can function as a theory in empirical science only if it has been given an *interpretation* by reference to empirical phenomena" (op cit, page 184). We find ourselves caught between a need to examine the interface between theory and empirical evidence on the one hand, and being limited in our attempts to construct theory by not being able to define clearly the systematic connections between observables and unobservables on the other. Perhaps some consolation can be derived from Mario Bunge who suggested (1967, page 397):

"... by and large, theorems come in time before axioms, even though axioms are logically prior to theorems. The historical pathway has almost always been the following: (i) establishing new level generalizations (future theorems); (ii) generalizing the foregoing; (iii) discovering logical relations among known theorems; (iv) 'discovering' that some of them might serve as axioms or inventing higher-level premises out of which the available body of knowledge can be derived."

What seems to be true of the social and behavioral sciences in general has been a procedure of trying to produce theories first and worrying later about exposing their logical skeletons. In geography to date, such exposures have generally been discouraging.

Thus, problems faced in theory construction and model building for human geographic research are similar to those facing social scientists generally. Two of the most fundamental problems are those of defining the domain of a theory and identifying its 'universe of discourse' (or the time-independence of its relations). Other common problems include operationalizing concepts used in the theory and models derived from it, and formalizing the direction and nature of the logical deductive nets included in the theory.

Problems peculiar to researchers attempting to construct theory in geography arise from the fact that they must be interested not only in the external physical environment and the internalizing of human actions, but also with the interface between the two. This introduces problems of how to represent cognitive and physical worlds, how to incorporate societal and institutional constraints, and how to use behavioral processes to explain overt activity in the physical world. Finding answers to such questions seems to be an enormously difficult task, and the search for solutions appears to define conflicts that are causing further fragmentation (diversity?) and contributing to the confusion (richness?) dominating human geography today.

Conflicts inherent in the search for "Common Ground"

My first reaction to writing an essay for the conference which spawned this book was to write a short statement under the title "Where Have All the Dreamers Gone?" If one takes a look at the series of books and papers that influenced the discipline in the 1960s, the most obvious thing about each of them is that they are incomplete, speculative, and fragmented! Some, although *written* with considerable clarity, are without a doubt unclear about the message they wish to impart. Much of the written work in geography did not solve problems but raised questions. Many authors suggested a possible answer or an approach to solving the questions they raised, but without doubt the most significant contribution of their work was the *question asked*, not the procedures used or the solutions obtained. The people who wrote these were the dreamers; they were the individuals who saw a future, a way for the discipline to go. They were the ones who saw a void in existing knowledge that had to be filled, who directed researchers down a path of excitement to the gate of knowledge. Where have all the dreamers gone? Some, like Gunnar Olsson, have followed their own path, have knocked lustily at the gate of knowledge, and have freed their intellect and their spirit in following the path they chose. Others, like Peter Gould, were hungry for experience and knowledge and constantly sought to improve the mechanisms that allowed them to accumulate awareness and translate it into knowledge. Many became obsessed with seeking solutions, and for the next decade they wrote papers that merely solved problems. Many of the problems were trivial, and by the time they had been solved, were of little interest.

Herein lies the first of the conflicts that have become obvious in the discipline, and which are now seen to prohibit the attainment of a common ground. As geography has changed its image from a 'soft' to a 'hard' subject area, it has become much more difficult for the dreamers. Now everyone is aware of the need for precision and accuracy in their statements, of the dangers of leaving comments unsupported (that is, of speculating), of leaving themselves open to charges of reductionism, and of using polemic and rhetoric instead of entering into a critical dialogue. The Dr Jekyll in us wants to be objective, general, concrete, factual, holistic, empirical, pragmatic, and certain. But Mr Hyde insists that we be subjective, particular, abstract, ambiguous, belief-laden, fragmented, and uncertain. Our Dr Jekyll wants to continue the scientific modes of thought that allowed him to take a comfortable place as a well-paid member of society. But Mr Hyde's humanist predilections force the rejection of those values, and even a rejection of the society which gives a paramount position to such values. The geography profession currently is uncertain as to which personality is the evil one, and this schism exists as a fundamental deterrent to the acceptance of a common ground on any basis except its use as the site for battle.

The scientific–humanist controversy is not the only conflict in the discipline. There is continuing division over such things as the relative value of theory; the importance of practical applications of knowledge; the worth of process-driven rather than cross-sectional explanations; the significance of micro-level rather than macro-level approaches; the importance of human behavioral processes and the role of the individual in spatial systems; the relative significance of institutional and societal constraints on human actions; the selection of radical as opposed to conventional points of view as the bases for interpretive schema; and, of course, the time-honored dichotomy between 'physical' and 'human' geographers. Each of us can add to this list; but given such a state of affairs, what hope is there for finding a common ground on which to base our search for knowledge?

Perhaps much of the confusion that lies at the heart of geography today results from an awareness that there are simply many geographies and many possible worlds. Uncertainty arises because we know not which geography to choose, nor which possible world we should aim for. We run the risk of becoming dogmatic by trying to force all worlds into one very limited format, and in doing so we ignore, belittle, or forget the others. We were suspicious of the world of rational economic man, but this suspicion helped us both to comprehend it and to change it. But economic man was an integral part of much spatial theory, and in rejecting him we have emasculated those theories in turn. It is imperative that in our search for better theory, and for a better understanding of spatial phenomena, we are not trapped in the equally abhorrent world of Marxist man or phenomenological man, or psychologist man, or political man, or traffic engineer's man, or national statistician's man, or townplanner's man —or any of the others that Helen Couclelis has so clearly listed for us (1982, this volume).

In the search for understanding, in our quest for knowledge, we must also be careful not to accept the exhortations of those narrow-minded few who urge us to dispense with all prior knowledge, to start from a 'clean slate', to defy the establishment and become anarchistic. We cannot afford to throw away what little knowledge we have so grudgingly and so painstakingly extorted from the flux of everyday existence. We may at this time be uncertain as to how we can best use the knowledge we have obtained, but let us not dispense with it out of hand. Helen Couclelis echoes my own thoughts in the first page of her essay (page 105, this volume) when she says, "Fragmentation of research is not necessarily in itself an evil: the world is after all a very varied place, and we need all the different perspectives we can get." She further suggests that she would rather see some blows of antagonism exchanged rather than subscribe to dogmatism, and I must heartily agree with this position. It reflects a fundamental premise of difference that is essential for humanity. Let each of us stress our individuality and our difference, our temperaments, our beliefs, our knowledge levels, our interests, our perspectives.

Helen Couclelis also expresses my sentiments exactly when she says, "My conception of a successful geography is of a discipline which does not turn to what philosophy is available for seeking guidance from above, but is *inherently philosophical as much as it is technical, eventually becoming itself a source of new material for philosophy in the way the most advanced branches of theoretical physics have done*" (page 108, this volume). The question raised by this statement is a critical one. It touches on our belief in the worth of geographical research, even as it touches on the root of all knowledge. Are geographers doing anything of any significance? Couclelis looks forward to the day when professional philosophers will have a conference on 'philosophy in geography'. I would settle for much less than this, and be happy to see at first (on an intermittent and then on a regular basis) conferences in other disciplines on what geography can contribute to their own body of knowledge.

References
Ackerman E A, 1963 "Where is a research frontier?" *Annals of the Association of American Geographers* 53 429–440
Bergmann G, 1958 *Philosophy of Science* (University of Wisconsin Press, Madison)
Berry B J L, Simmons J R, Tennant R J, 1963 "Urban population densities: structure and change" *Geographical Review* 53 395–405
Brown R, 1963 *Explanation in Social Science* (Aldine, Chicago)
Bunge M, 1967 *Scientific Research I: The Search for System* (Springer, New York)

Couclelis H, 1982 "Philosophy in the construction of geographic reality" in *A Search for Common Ground* Eds P Gould, G Olsson (Pion, London) pp 105-138

Golledge R G, 1973 "Some issues related to the search for geographical knowledge" *Antipode* **52** 60-65

Golledge R G, Amedeo D, 1968 "On laws in geography" *Annals of the Association of American Geographers* **58** 760-774

Gould P, 1979 "Signals in the noise" in *Philosophy in Geography* Eds S Gale, G Olsson (Reidel, Boston) pp 121-154

Gould P, 1980 "A search for common ground" (position paper for the Conference on Philosophy in Geography, Bellagio, Italy)

Hempel C G, 1965 *Aspects of Scientific Explanation* (Macmillan, New York)

Lee H N, 1968 "Conceptual models in knowledge" *Tulaine Studies in Philosophy* **17** 101-113

Lee H N, 1973 *Percepts, Concepts, and Theoretic Knowledge* (Memphis State University Press, Memphis)

Rapoport A, 1959 "Uses and limitations of mathematical models in social science" in *Symposium on Sociological Theory* Ed. L Gross (Row and Peterson, Evanston, Ill.) pp 348-372

Geography: paradigmatic change or functional recovery? A view from West Germany

Dietrich Bartels

Views and opinions of geography—the hidden self-contempt

When you ask geographers why they *themselves* originally chose their
subject, and continue today in their professional geographic jobs, and then
you ask them why *other* colleagues do so, you frequently find that they
give quite honorable reasons for their own preferences, but ascribe rather
suspect motives to their professional peers. Moreover, when you ask
them how they think nongeographers would comment on geography, they
usually expect to hear a preponderance of rather adverse judgements.
Thus, the inferior opinion of one's professional self is often concealed and
projected onto the putative judgement of other people. In other words,
the so-called miserable image of our discipline is really a feeling of
inferiority characteristic of many geographers themselves, and as such it
appears to be widespread—even though people may not willingly talk
about it in public. Yet it is precisely from this situation that a permanent
attitude of vindication and opportunism arises against a background of
repressed motives (cf Hard, 1979).

Conversely, if we interview nongeographers from all professions and
social classes for their assessment of geography, we usually find that they
make two, rather clear distinctions. First of all, they frequently comment
upon the dull and undemanding geography lessons they received in school
when they were children, but then they go on to contrast these experiences
with their own, quite clear appreciation that geography can make valuable
contributions to general education as a subject of distinct importance in
the formation of human culture, and even as a topic helping to orient a
person in the conduct of his or her life. This sort of quite deep
appreciation for the subject appears to surface mainly when people in
other university disciplines are questioned and asked to give their general
appraisal of the subject. Unfortunately, such general goodwill towards the
discipline is greatly diminished when scholars in other fields with strong
traditions of exact, empirical research are questioned about the depth and
quality of geographic inquiry. In brief, despite the discipline's own belief
in an unfavorable view of itself from the outside, geography, considered as
a general subject worth teaching, is quite popular and well regarded,
except as a stringent research discipline. Yet this raises a distinct problem
and question: in what sense or function can we say that geography is
genuinely well regarded? What does it mean that we are regarded
favorably as a worthwhile subject to teach, yet are most unfavorably

appraised for the very process of research that should lead to expanded geographical knowledge and understanding?

The real genesis of (German) university geography

I cannot speak for other countries, but certainly in Germany the expansion of geography out of the traditional domain of ordinary school education was encouraged by the expanding intellectual interests of the Bourgeois during the 19th century, interests which were nourished in turn by the accounts of individual savants and explorers. In fact, the expansion of the subject into the sphere of university education has largely taken place against the advice, and even the outright opposition, of the traditional faculties. Far from being invited into the university, geography was essentially forced upon it by governmental authorities who hoped to see an effective and school-oriented new science for the training of teachers, as well as a subject enhancing general national education. The latter goal was interpreted as the formation of a respectiable and reliable view of the world, seen in a patriotic-nationalistic perspective. Those who held the new professional chairs of geography at German universities were consistently people who previously had been school masters—autodidacts with rather polyhistoric interests, and with very varied educational qualifications that frequently placed them somewhere between the philologist and the apothecary. They had, in fact, no contacts with the former generation of famous world explorers, nor with the contemporary world of commerce after 1880, two aspects of the time that were both oriented in actual practice to the discovery and exploitation of colonial resources overseas. Thus the scientific beginnings of the higher academic form of geography simply cannot be compared to the intellectual and international standing of other disciplines in German universities around the turn of the century.

Given this particular, not to say unusual beginning in the university, the first aim and policy was to provide more geography lessons in the schools, and better training for teachers of the subject at the school level. It is for this reason that school-directed regional knowledge has formed the main content of the curriculum, as well as the driving force of the discipline's 'progress'. It should come as no surprise that in the eyes of the already well-established university disciplines geography hardly qualified as a subject for real and profound research. As a matter of fact, most of the contributions to the *Geographisches Jahrbuch* over seven decades (1866–1937) were written by geologists, geophysicists, botanists, zoologists, statisticians, or ethnologists. In contrast, the geographers, as the real editors, produced only methodological and review supplements. Thus it was not long before the need for a second policy arose in the discipline, a need to create a conception of a reputable general science of the world's surface, a conception bolstered by the myth that there always had been a time-honored science of geography, which had given birth to, endowed,

and reared most of the other branches of research—often at the tragic
cost of important parts of its own original subject. This phantom and
mythical idea of geography as a serious and stringent empirical discipline,
with 'proof' of a long tradition of true and deep research, still forms part
of our heritage today.

The struggle for respect as a true research discipline
For over a century, geography has been competing with other university
disciplines for public recognition—not to say public funds. During the
course of this century, several striking attempts were made to identify the
'true essence' of geographical research. These attempts have usually been
centered around either a specific aspect of theory building, or on a
particular metatheoretical position. For example:
(a) The oldest paradigm has focused upon the systematic investigation of
reciprocal relations and interactions between *'Man and Environment'*,
an approach for which, until quite recently, no consistent theory of
systems had been constructed, and an approach which still produces in
Germany only trivial findings—in spite of the current widespread
environmental concern, and in contrast to the later and continuing
application of general systems theory to problems in geomedicine,
bioecology, and other disciplines.
(b) Another tradition has appealed to *'Länder'* ('regions') as unities, or to
'Landschaften' (landscapes) as totalities, both of which have been
considered as autonomous empirical subjects worthy of their own
specialized discipline. Such an appeal was enunciated in its most
sophisticated form by Hettner and Schlüter during the years between
the two world wars. In this same period, specific research methods
were also often referred to (although with a rather different stress),
methods which were considered to be the only suitable ones to elucidate
such special objects of study—the peculiarity of landscape observation,
the special 'synthetic' method of geographical understanding, the
approach of 'historical' explanation, and so on. Certain bases of such
special methodologies can be identified in some of the philosophies in
vogue in German-speaking countries between 1890 and 1930—for
example, phenomenology and idealism.
(c) A more recent perspective might be termed the *'quantitative revolution'*,
a paradigm rather hesitantly adopted from the English-speaking world
after the 1950s. In Germany the discussion of this horizon has been
characterized not so much by the struggle to establish quantitative
techniques (which even now are regarded by many as dangerous or
bewitching), as by the demands for the formation of explicit theories
under the banner of positivistic-nomological ideals for a true science in
the sense of empirical and analytic philosophy. However, despite the
common conviction of all academic geographers that they share the
strong desire to prove the true research character of geography, the two

camps of advocates and antagonists of the quantitative revolution opposed each other very strongly. In fact some geographers, who wanted to be considered as genuine scientists, refused this new kind of theoretical rescue attempt, while others overemphasized the intellectual carrying capacity of the new quantitative platforms. Surprisingly, this dogmatic battle has been dropped since 1975, without acknowledging any decisive result, whereas factually quantitative methods have entered silently and continuously into the fabric of German geography.

(d) A fourth attempt during the 1960s was founded upon the discovery or renaissance of the specific conception of a *science of distance and space* being considered as a new unifying force among geographers, particularly appropriate for its ability to keep human and physical geography from drifting too far apart. However, the narrow basis of the conception was highlighted very quickly. One of the problems was that this perspective was hardly the exclusive preserve of geography: other disciplines had also developed research perspectives focusing explicitly upon problems of spatial model building. A second problem was that the various sorts of earth distances used by geographers as explanatory variables seemed to become less and less important. In fact, they often lose their value as empirical variables within the framework of relevant and acute research approaches, and so become merely indirect and imprecise indicators (for example, 'distance decays') of the 'true' mechanisms of spatial transmission or resistance between subjects of all fields of physical or social theory. Besides, physical geography never really felt obliged to adopt the spatial approach as its own broad framework for research.

(e) A final attempt to anchor geography is indicated by the so-called *behavioral approach*—a sort of reinvention of psychology within geography. This research perspective arose and spread during the 1970s, and in a sense tried to gather together all the ancestral man–environment problems of geography which appear to be threatened by competitive disciplines. These are to be saved for geography by bringing them into the subjective realm of individual perceptions, preferences, values, and motivations as they related to human decisions and actions. The importance of such subjectivity is often extolled as the new and vital research task of human geography, which includes nature as well as many man-created subjects as substrata of 'spatial perception'.

The struggle for the social function of academic geography

There is no question that scientific activities and results have a societal function—at least in the long run. Even 'ivory towers' are subject from time to time to queries from the outside, not the least of which are proofs of their existence, the worth of their research findings, and their pedagogic relevance. In Germany, these sorts of queries and challenges to academic fields have broadly emerged since 1945, and geography as a discipline did not appear to be particularly relevant during the overall national reconstruction phase, although whatever small importance it did have appeared to arise from its image as a discipline that was in some way *engaged*. However, geography as an engaged discipline was viewed from two very different perspectives, characterizing two diametrically opposed camps.

The view from *one* camp is that scientific findings should be transformed directly into instructions for social action and into technologies for planning progress. Since geography is in competition with all the other disciplines, and is also under the same sort of public pressure, it should legitimate itself as an *applied discipline* by contributing to the solution of some of the 'urgent' problems of social engineering. Thus in particular, it should try to improve the spatial order of society, and do this by cooperating as much as possible with public authorities—who would then lend, in turn, some of their official prestige to the engaged geographers and their universities.

However, what we have typically seen, after the rather desperate attempts to prove the legitimacy of geography in the 1950s, is that the more advanced and comparatively efficient praxis-oriented fields within academic geography have tended to disengage themselves from the rest of the discipline. Within the universities themselves, regional science joined forces with some geographers, economics, city planning, and the engineering sciences as part of the larger effort to generate interdisciplinary planning and policymaking studies. Outside of the universities, competent professional geographers are to be found in specialized advisory institutions and research and development companies, and although they can hardly be identified purely as geographers anymore, they are frequently noted with pride by many back-slapping university colleagues.

This somewhat opportunistic view of applied geography as a discipline assisting public authorities wherever a possible application could be identified, or whenever contacts with respective authorities could be forged, has been superseded today by the *moral appeal* to help 'minorities' of all kinds—miserable suburbs, retarded regions, and underdeveloped countries. Geographical investigations focusing upon this kind of problem would help to overcome human *disparities* of all sorts, creating a better world grounded in the moral tradition of Western humanism. In actual fact, German geography has made few really practical efforts to follow such a moral appeal, although applications of new sorts of models are increasing, rooted in the development of regional disequilibrium or pole theories on all scales, and they contrast quite markedly with the equilibrium and free-market models of the types of von Thünen or Christaller.

The view from the *other* camp is represented by the demand to concentrate upon the intentional or involuntary influences of scientific activities and theories themselves on society, particularly as these have been accentuated by the tradition of *critical theory*, above all of the

Frankfurt School. In geography, this has meant a reconsideration of the basic paradigms of the discipline, recognizing the role of geography as one of many ideological instruments for structuring a society or stabilizing its state. Thus the old concepts of *Landschaft* and *Totalcharakter eines Landes* (the totality of a region) are 'unmasked' as vehicles of general societal education for the purpose of making people internalize perspectives of the world which are particularly appropriate for the larger societal system. Thus bourgeois social geography may appear merely as a rather bland and superficial description of the deeper structure of repressive and forced behavior, and far from the dynamic subject that might be appropriately taught to emancipated human beings in an ideal society. Similarly, the usual regional geographies of foreign countries are either viewed as instruction that supports and affirms capitalism and its impacts, or as possible intellectual lanterns that could cast a bright light on the realities of imperialism overseas. Whatever the case, scientific theories and inquiries in geography should have their direct ideological functions in order to expose the miserable condition of present society, a condition which must change completely in the near future according to more or less radical new models of man.

Somewhat ironically, there has been a growing awareness that critical theory can only be as critical and as harsh as it has been if it holds to some ideal model of man and society. This model is itself a result of a partial or biased ideology, and no matter what dialectical interpretations it invokes, it can never sail outside of the circular problems of hermeneutic interpretation.

Reconsideration of geography as a nonresearch discipline

Since 1975, most academic geographers in Germany have adopted an attitude of what we might call 'privatism' when it comes to problems about the basic foundations of the discipline. There is almost no real debate today—certainly nothing resembling open conflicts of opinion—and most scholars of geography seem to be occupied only with their own isolated and quite individual research 'hobbies'. Nevertheless, I have the distinct feeling that a countertrend is developing, a desire to discuss once again questions about the foundations of the subject. This comparatively young movement of self-interpretation within German geography has at least three roots or origins of different ages.

First, and since the 1960s, there have been two distinct trends in the development of the intellectual life of modern, postindustrial societies. We have witnessed the growing degree of specialization of individual research disciplines, especially those which correspond closely to the ideals of positivistic–nomological theory, and we have also seen the growing complexity of the numerous and unavoidable practical decisions faced by individuals and social institutions in their respective life environment. The trend to narrow specialization, in contrast to the growing complexity and interrelatedness of decisionmaking, has created an enormous gap between deep and specialized knowledge and the requirements for higher inter-connected decisions. It is for this reason that it has become absolutely indispensable to create *interfering institutions* for mediating, combining, and processing scientific knowledge before it can be applied and used, and for structuring human problems before they can be solved by scientific inquiries—if they can be solved at all. Businesses and large industries established their own agencies very early on to connect scientific research and marketing, and most departments involved in public administration also have their special boards for functional research related to decisions that are essentially political. But who carries out scientific research and guidance for the individual citizen? And what about the social status of all these advisory boards and their members as compared to the fairly prestigious research scholars in the universities? Do the latter only perform a 'feeder' service, or are they part of the main management?

Second, a great deal of thinking has been devoted to the essence and purpose of *school education* since the 1960s, reflections characterized by extensive debates on learning objectives (for example, the curriculum discussion based on Robinsohn). These discussions on various educational models, and what these mean in terms of reforming the school disciplines, have created a strong sense of the necessity to introduce youth into 'their own life' and not into 'isolated and abstract findings about life', that is, to guide young people to find their own individual ways in our increasingly more complicated and functionalized world. Given this fundamental purpose, a distinction is frequently made between two partial tasks: first, to instruct the individual as a *consumer* to satisfy the totality of his basic needs in an optimal manner; and, second, to inform him as a *citizen* on his possibilities for participating in common decisions and activities to improve his own society and environment.

The question thus arises: which school discipline could perform these basic tasks, and to what degree could it do it effectively? Which subject within the canon of classical school disciplines seems most predisposed and capable of undertaking such basic educational objectives? Faced by such crucial questions, German school geography (*Schulerdkunde*) suddenly remembered its former tradition (factual or imaginary) of introducing practical knowledge of its own and foreign countries and cultures, and began to refurbish its programmes accordingly to teach children about modern human environments—even going so far as to include certain aspects of critical and emancipatory teaching on human society. One example of this trend of renewing school geography, and reorienting it to all round instruction about human life, is the RCFP (*Raumwissen-schaftliches Curriculum-Projekt*; Geipel, 1975), which relied partly on the methodological experiences of the American HSGP (High School Geography Project).

These sorts of discussions were brought to a close around 1972. For
the most part, they had been undertaken by people outside of the
universities themselves, mainly because those in the universities appeared
to have not the slightest desire to reorient themselves to such 'humble'
didactic problems. These generally had the intellectual, not to say 'snob'
appeal of independent research in a somewhat 'higher' and autonomous
academic science. Not surprisingly, this rebuff induced school geography
to seek its new intellectual equipment and support from other pedagogic
institutions, or to look for advice and scientific input from other more
relevant and 'true' research disciplines—instead of merely university
geography.

Third, there is a growing awareness that increasing technological
possibilities endanger civilization by creating new ways of manipulating the
world, and that these dangers are intensified by the decreasing efficacy of
public management. Public administrators are torn between the
requirements of consistent and centralized coordination of a large number
of interlocked components of government, and the demands for
decentralized, democratic, and local control by people over their 'own'
small group environments. These extraordinary tensions require that we
develop new ways to cope simultaneously with an overall or total view of
all needs of life of *all* individuals in *all* environmental situations.

One variant of this call for balancing the global view with local desires
brings to the fore spatial aspects of genuine importance. These are
summed up in what we might call modern *Regionalism*—a concept which
in Germany (as in the rest of Western Europe), is pleaded for by the
'Greenearthers' (*die Grünen*), and their political party formations. These
people argue for a reduction in what they believe to be a much too highly
developed spatial division of labor both at the national and at the world
scales. They believe that such extreme specialization results in too much
economic power and concentration, and they would prefer to accentuate
lower levels of economic activity within local units or cells—small regions
approaching the autonomy of a canton and offering 'true homelands' for
their respective populations, people who feel themselves unsheltered and
homeless today within a mass society disintegrated and overfunctionalized
from afar.

These sorts of perspectives on the creation and synthesis of small
autonomous 'lebensraums', each with its own individuality and uniqueness,
doubtless have some precursors in the ideas of total regions of classical
regional geography. To the extent that these intellectual trends behind
regionalism parallel the present demands of the Greenearthers and become
a general background of a majority, so geography might find a new
opportunity to evolve towards a discipline of life support and orientation,
contributing directly to new needs of society and its individuals. However,
and as a direct result of such a trend, academic geography might move
away from the objectives of exact or 'positivistic' research (a trend

completely consistent with the general antiscientific feelings shaping regionalism and related trends), and so return to yet another descriptive regional approach. Unfortunately this would lead only to a new trivialization of the field.

A new period of clarifying objectives

Quite independently of the two future scenarios I have outlined above, we shall have to reflect deeply upon the role of geography within society. We have to face the new and critical judgements about positivistic research ideals, with all the growing doubts about problems of correspondence between theoretical terms and empirically operationalized definitions, about critical theory and its hermeneutic boundaries, and about the present situation of our societies and their needs and demands for science. An intensified 'search for common ground' among academic geographers has begun, and I wish to close with some arguments from the present German discussion.

First, and as I noted at the beginning of the essay, we have the opportunity to revitalize one of the old and genuine virtues of geography, particularly as it is seen by outsiders. This is the privilege of bringing new phenomena into the field, with the freedom to raise new questions about them. We have the opportunity of developing new and audacious approaches towards the problems of tomorrow, without feeling constantly obliged to furnish exact proofs or immediate solutions, and without worrying all the time about where the boundaries of adjacent disciplines lie. There is no question that we need new ideas about the interconnected structure of things in our human world, and it seems quite appropriate to work towards what we might call a 'new Chinese atmosphere', where many flowers bloom and multiple paradigms are tested within geography at the same time.

Second, there are undoubtedly valuable aspects and methods within modern phenomenology which might augment, complement, and correct some of our current ways of acquiring empirical knowledge. These might be especially valuable as we try to gain better insights into what has been labeled the 'totality' of our life situation. Given certain, and rather extreme trends towards abstraction, it can certainly do no harm to follow Husserl and go "back to the things themselves" as a complementary perspective.

Third, we have the opportunity to contribute strongly to a general programme of education to enhance public awareness that the future qualities of human life are the main and *normative* problems of our society. Geography could adopt the stance of a general didactical discipline, with many of its members playing the role of social advocates outside of any particular lobby, pressure group, or public institutional framework. This might even emerge as a new definition of geography!

Last, we should consider the question of how to formulate and isolate partial, but well-defined, problems within the whole context of man and the environment, indicating those areas of expertise that could solve them, and trying to bring about a sense of cooperation between existing research disciplines. This is *not* the classical image of the synthesizing musical director, conducting an orchestra of other disciplines, but a sort of counterpart to all the positive sciences, an intermediating factor between *Erkenntnis* and *Interesse* to serve human needs as a humanistic connecting link.

References
Geipel R, 1975 "Das Raumwissenschaftliche Curriculum-Forschungsprojekt" in *Der Erdkundeunterricht*, Sonderheft 3, Stuttgart
Hard G, 1979 "Die Disziplin der Weißwäscher. Über Genese und Funktion des Opportunismus in der Geographie" *Osnabrücker Studien zur Geographie 2: Zur Situation der deutschen Geographie 10 Jahre nach Kiel*, Eds G Hard, H-J Wenzel (Universität Osnabrück Neuer Graben/Schloss, Osnabrück) pp 11-58
Hettner A, 1927 *Die Geographie, ihre Geschichte, ihr Wesen und ihre Methoden* (Hirt-Verlag, Breslau)
Husserl E, 1950 *Die Idee der Phänomenologie: Fünf Vorlesungen* (G Grote, Den Haag)
Robinsohn S B, 1976 *Bildungsreform als Revision des Curriculums* (Neuwied, Berlin)
Schlüter O, 1928 "Die analytische Geographie der Kulturlandschaft" in *Zeitschrift der Gesellschaft für Erdkunde zu Berlin*, Sonderband, 388-411

Clarification and meaning

Geography as a human science: a philosophic critique of the positivist-humanist split

Kathleen Christensen

A human science differs, first and foremost, from a physical science because its subject matter is *man*. As agent, man interprets and gives meaning to his everyday actions and relations with other men and things. His interpretations and meanings constitute his lived world[1], and they exist prior to, and independent of, any scientific explanations of them. It is the lived world of man that provides the proper subject matter for a human science.

In contrast, the subject matter of a physical science consists of non-human things that do not have the capacity to interpret themselves or the world in which they exist—a rock has no meaning for another rock, and it does not have the capacity to know a 'rock' world. This crucial difference in subject matter presents a unique challenge to those working in a human science. Unlike the physical scientist, who concerns himself only with meanings that he himself projects onto things, a person working in the realm of the human sciences must take into account the preinterpretations by man of his lived world. In effect, a human science must contend with two orders of meaning—the scientific and the lived—because a scientific theoretical framework exists as the second order of meaning which is constructed and derived from the lived world as the first order of meaning (Schutz, 1962, pages 5–6). The current split in human geography along

---

[1] A world is a finite province of meaning, and the issue of multiple worlds has been discussed and examined by several phenomenologists, including, but not limited to, Husserl, Heidegger, Gurwitsch, Schutz, and Kockelmans. Husserl concerned himself with the issue in *Ideas* (1975) and *The Crisis of European Sciences* (1970). In the latter work, he most thoroughly discussed the issue of *Lebenswelt*, the lived world. Heidegger examined the different modes of Being in the world in *Being and Time* (1962). Gurwitsch, somewhat along Husserlian lines, developed the issue of world in the context of the natural sciences. Kockelmans in *The World in Science and Philosophy* (1969) drew on Heidegger in examining the world as it is presupposed in the natural sciences and in philosophy. Yet it is in Schutz's work (1962) that we find the clearest discussion of world as it is oriented to the human sciences. In the chapter "On Multiple Realities", Schutz describes the world of daily life, explains its social structure, and relates this world to others, including the world of scientific theorizing (Schutz, 1962, pages 207–259). Although Schutz himself most explicitly attributes the direction of his thinking to the ideas of Husserl, Kockelmans (1979) has shown that, in the context of the phenomenological movement, Schutz's ideas align more comfortably with the tradition of Heidegger than with the tradition of Husserl.

Any ideas on world presented in this essay are derived, directly, from Schutz's work on world. Yet the ultimate source of inspiration for the ideas lies with Heidegger, even though his works are not explicitly or consistently introduced and referred to.

the lines of positivism and humanism obscures the meaning of geography as a human science, because neither interpretation can reveal how the two orders of meaning connect.

Positivism takes the lived world for granted, and fails to recognize or accept that this world actually constitutes the *foundation* of a human science. Rather, this interpretation of science focuses exclusively on the logical, linguistic, and methodological structures of the concepts, principles, and methods of the theoretical frameworks of a particular science. In so doing, it legitimizes only the world of science. In reaction to positivism, humanism pays attention only to the lived world and, in effect, ignores the scientific. Thus the positivist–humanist split severs the two worlds, and so masks the meaning of geography as a human science.

But the split cannot simply be pushed aside; it must also be placed in perspective. When positivism was explicitly introduced into the discipline in the middle 1950s, it gave the illusion—and at times the reality—of scientific rigor, because it held up as an ideal a scientific mode of inquiry that was monistic, reductionistic, and physicalistic (Radnitzky, 1970, page xvi). In fact, and to be strictly accurate, the ideal was one articulated by logical empiricism—a contemporary variant of positivism[2]. Human geographers inspired by the ideal were responsible for the analogical development and application of such physical models and theories as the gravity model, Monte Carlo simulation models, entropy maximization models, and catastrophe theory. As a result, they almost invariably concerned themselves with the second order of meaning captured in the concepts, principles, and language of the theoretical frameworks of physics. They arbitrarily superimposed such frameworks onto human phenomena, thereby reducing them to the concepts and languages of the physical sciences. Almost inevitably, humanism emerged in the 1970s as a reaction to the unreflective and arbitrary use of models and theories from the physical sciences.

Humanism, which tends to draw on a rather particular and sometimes neoromantic interpretation of phenomenology, rejects physics outright as an ideal model for a human science. However, in their efforts to reject positivism, the geographers of the humanistic tradition appear at times to reject empirical science itself. Intentionally or not, these geographers ignore the second order of meaning in their efforts to reveal the nature and significance of the prescientific lived world of man. Despite this oversight,

[2] Unless otherwise noted, when I speak of positivism I am referring to its contemporary variant, logical empiricism (logical positivism) which articulates a certain ideal of science. According to the ideal, all sciences should form part of one science (monism), and amongst the existent sciences, physics best approximates such an ideal (physicalism). Sciences other than physics should be reduced to and founded on physics, and they can be so reduced "by making a language that has been designed for an idealized physical science the common language of all 'science-like' disciplines and by making physical concepts the fundamental concepts of the ideal unified science (reductionism)" (Radnitzky, 1970, page xvi).

humanism should not be lightly dismissed, because it has brought to light the primary meaning of the lived world which has been buried under the weight of the theoretical *a priori* frameworks of the physical sciences. Positivism spotlights science, and humanism illuminates the lived world; yet neither interpretation provides insights into the possible connections between the two worlds. Such insights will reveal themselves only when the philosophical polemics which fuel the positivist–humanist split are subjected to a dispassionate critique. Such a critique requires that we examine with the utmost care the geographic claims regarding positivism, empirical science (that is, positive science)[3], and phenomenology.

Note, however, that a *critique* differs from *criticism*, and the distinction between the two is crucial. Criticism involves passing judgements, and it centers on finding fault. In contrast, the Kantian sense of critique involves laying out and describing precisely the claims being made, and then evaluating such claims in terms of their original meanings. In effect, the objective of a critique is to establish the limits of validity for the claims. In this essay, the claims will be examined in light of the works of Kant (1965), Comte (1864), Husserl (1975; 1977), and Heidegger (1962). The critique begins with the geographic claims about phenomenology, because it was the introduction of these claims that explicitly triggered the positivist–humanist split.

A critique of geographic claims

The split between positivism and humanism derives from well-intentioned, but ambiguous and sometimes misleading, interpretations of the original philosophical formulations of phenomenology, positivism, and empirical science, as well as their relationships to one another. In particular, the relationship of phenomenology to empirical geography has been cloaked in ambiguity, an ambiguity which may be attributed in large measure to the pervasive confusion in the discipline between a positivistic interpretation of science and the meaning of the empirical character of that science.

At least three distinct interpretations of the relationship between phenomenology and empirical geography have emerged during the last twelve years (Relph, 1970; Mercer and Powell, 1972; Tuan, 1971; 1974; 1976; 1979; Buttimer, 1974; 1976; Entrikin, 1976; Gregory, 1978; Ley and Samuels, 1978; and Hay, 1979). Relph (1970) and Buttimer (1974; 1976) characterize a dualistic relationship between *phenomenology and geography*, in which phenomenology provides an alternative or substitute method to the scientific method of hypothesis testing and theory building. The phenomenological method is used to reveal the 'subjective' experience of the agents of their lifeworld. Entrikin (1976) interprets phenomenology as a *criticism of geography*, by which "geographers can be

[3] In the Kantian sense, a positive science constitutes what we would commonly refer to as an empirical science. Therefore, the terms "positive" and "empirical" are used interchangeably when referring to that component of science.

made more self-aware and cognizant of many of the hidden assumptions and implications of their methods and research" (1976, page 632). Gregory (1978), in a somewhat different vein, takes phenomenology as a *critique of the positivistic interpretation of geography*. Of the three views, the view held by Relph and by Buttimer represents the humanistic position which attempts to penetrate to the primary lived world of man and so reveal its meaning independent of the scientific method.

Relph characterizes the nature of phenomenology as consisting of three basic elements (1970, page 193):

"First, the importance of man's 'lived world' of experience; second, an opposition to the 'dictatorship and absolutism of scientific thought over other forms of thinking'; and third, an attempt to formulate some alternative method of investigation to that of hypothesis testing and the development of theory."

Basically, he wants the phenomenological method to provide "a procedure for describing the everyday world of man's immediate experience, including his actions, memories, fantasies, and perceptions," and, thereby, to "provide a means of investigation through which the lived world of man's experience can be restored to a place of prominence in our thinking (1970, pages 193–194). With the phenomenological method, he can reject "the approaches of a mechanistic science and the search for scientific laws which have no meaning for man" (1970, page 194). Relph does not portray phenomenology as an "irrational anti-science", but rather as a position in opposition to the "assumptions and methods of physical science, especially those of positivistic science and scientism" (1970, page 195).

Buttimer follows Relph's line of interpretation, but emphasizes the role of human values in phenomenology. She claims that phenomenology, "unlike other philosophical traditions in the West which have tended to separate 'values' from 'facts', and the values of actions from the agents of those actions ... considers values as central to the whole human experience" (1974, page 37). Accordingly, she characterizes the prime objective of phenomenology to be "the exploration and understanding of meaning and value" (1974, page 37). In her subsequent work, she continues to emphasize the role that the phenomenological method plays in elucidating the meanings that agents attribute to their actions in their respective life-worlds (1976, page 277). However, she differs from Relph, in statement at least, in that she does not see phenomenology as a method for revealing the individual, subjective experience of an agent, but rather for revealing the intersubjectively shared meanings of the agents in their lifeworld and of the scientists studying these agents (1976, page 291).

Although Buttimer does not pit phenomenology against science, she views it as a means to challenge "many of the premises and procedures of positive science", such as reductionism, (instrumental) rationality, and the separation of subject and object (1976, page 278). She does not believe that phenomenological inquiry and scientific inquiry are inevitably opposed: on the contrary, she argues that "we need to find their appropriate roles in the exploration of human experience" (1976, page 290). Presently, however, she can only maintain that "it is in the spirit of the phenomenological purpose, then, rather than in the practice of phenomenological procedures that one finds direction" (1976, page 280). Thus, when she concludes that "phenomenology invites us to explore some of the unifying conditions and forces in the human experience of the world" (1976, page 280), she speaks for all humanistic geographers who stress the integrity of the primary lived world, and who call attention to the geographic need to reveal and examine this lived world. At the same time, these particular humanists fail to lay out the structural connection between this primary lived order of meaning and the secondary scientific order of meaning. Entrikin (1976) and Gregory (1978) both attempt to provide an interpretation of phenomenology that reveals this connection, but they differ from each other in significantly fundamental ways.

Entrikin (1976) reviewed the humanistic literature in light of the phenomenological literature upon which it drew. He characterized the humanistic efforts as attempts to provide an "alternative to, or a presuppositionless basis for, scientific geography" (1976, page 615). He argues against such a characterization, maintaining that the phenomenologically-inspired humanism can be best understood as criticism (1976, page 632):

"As criticism it provides a potentially useful function in reaffirming the importance of the study of meaning and value in human geography, making geographers aware of their often extreme interpretations of science, and making scientists aware of the social and cultural factors involved in so-called objective research. As criticism, however, the humanist perspective does not fulfill the role suggested by some of its proponents of providing the essential insight, or presuppositionless basis for, a scientific geography. Humanistic geography as criticism is one of a number of means by which geographers can be made more self-aware and cognizant of many of the hidden assumptions and implications of their methods and research."

Gregory differs from Entrikin when he characterizes phenomenology as a *critique of positivism*: "Where Husserl's work has been of particular importance, therefore, is in its vigorous rejection of the aspirations and assumptions of positivism" (1978, page 130). Such an interpretation leads Gregory to view geographic efforts with phenomenology as efforts "directed towards the destruction of positivism as *philosophy*, rather than the construction of a phenomenologically sound *geography*" (1978, pages 125–126).

The differences between Entrikin's and Gregory's interpretations are twofold: it is the difference between criticism and critique; and it is the difference between what is the subject matter for criticism or critique.

As noted earlier, criticism is directed at finding fault with, whereas critique involves laying out the claims being made and evaluating them in terms of their genuine meaning. According to Entrikin, phenomenology subjects the empirical work of a discipline to criticism. In effect, he maintains that phenomenology "finds fault" with empirical research. Gregory, on the other hand, interprets phenomenology as subjecting the positivistic claims regarding science, such as reductionism, monism, or physicalism, to a critique.

These two interpretations point to radically different potentials for the phenomenologically-inspired humanistic branch of geography. If one follows Entrikin's interpretation, humanistic geography delivers a fault-finding criticism of empirical research, whereas if one follows Gregory's interpretation, the same humanistic perspective presents a clarification and critique of the positivistic interpretation of science. The basis for resolving the differences among the three interpretations rests in philosophy and the original philosophical formulations of positivism, empirical science, and phenomenology. But before we turn to philosophy, I feel we must attempt to uncover the factors which have led to these three different interpretations, since they appear to have precipitated and rigidified the positivist–humanist split.

There appear to be at least two significant factors which have contributed to the three diversely ambiguous interpretations of phenomenology's relationship to geography. First, the humanistic geographers—those who see phenomenology as a substitute or alternative method to the hypothesis testing of empirical science—have frequently drawn on a phenomenological literature that differs radically from Husserl's original formulation of the relationship of phenomenology to empirical science. Second, the humanists and the positivists alike have consistently confused positivism with empirical research. In doing so, they have implied that if one engages in empirical research, then, by definition, one is a positivist in the logical empiricist sense of a person who idealizes a human science in terms of physics. Both factors are deeply embedded in the geographic literature of the past ten years.

In 1970, Relph credited phenomenological psychology as the source of phenomenology in geography. The particular form of phenomenological psychology to which he refers has been promoted at Duquesne University, and it represents an attempt to provide a subfield or an alternative method to empirical psychology (Georgi et al, 1971). Subsequent humanistic geographers have continued to ground their empirical work in this particular interpretation of phenomenological psychology and they have used it in turn as a model for a phenomenological geography (Rowles, 1978; Seamon, 1979).

Unfortunately, what we might call the 'Duquesne interpretation' of phenomenological psychology constitutes a distinctly *non*-Husserlian conception of phenomenology itself[4]. Since Husserl is accepted as the father of the phenomenological movement, the difference is of fundamental significance. Any other geographic interpretations of phenomenology, such as those by Entrikin and Gregory which draw directly on Husserl, will be bound to differ in some ways from the interpretation held by humanistic geographers. The differences in the source materials on phenomenology will lead to ambiguously different interpretations of the relationship between phenomenology and empirical geography.

I must stress that I am not criticizing the Duquesne interpretation as ideas *in and of themselves*, nor am I denying that their ideas could be of importance to psychology, nor am I questioning whether their approach has led to important research in the human sciences, including human geography. All I am arguing is that what these particular 'phenomenological' psychologists are claiming is not what Husserl (1977) and Sartre (1948) had in mind when they spoke about the need to develop a phenomenological psychology[5]. Furthermore, they do the community of scholars a major disservice by referring to their own views as "phenomenological", a phrase and claim that suggest there are no further ideas in Husserl and Sartre that could be relevant to the human sciences.

Husserl and Sartre never intended phenomenology to represent an alternative to empirical science, nor did they ever see it as a subfield. They merely wanted to use phenomenology to provide a regional ontology as a foundation to an empirical science, because without such a foundation that science would be open to positivistic interpretation.

In effect, phenomenology constitutes a rejection of *positivism*, but not a rejection of a *positive*, that is to say *empirical*, science. But this point is lost in human geography, because human geographers have not consistently differentiated between positivism and empirical science. For example, Buttimer views phenomenology as challenging the "premises and procedures of positive science" (1976, page 278), whereas Gregory sees phenomenology as a critique of positivism (1978, page 278). A close reading of the two reveals that they are talking about the same phenomenon—a logical empiricist ideal of science—but are calling it different things. This confusion between positivism and positive science is exacerbated by geographic efforts to define positivism in terms of two positions regarding the structure of explanation in empirical research: Hempel's covering law of explanation; and Kuhn's paradigms of normal science.

Olsson characterizes Hempel's covering law of explanation as representing "the positivistic view of explanation and prediction" (1970, page 363).

[4] For a more thorough examination of the Husserlian conception of the relationship of phenomenology to psychology, see Husserl (1977) *Phenomenological Psychology*, and Sartre (1948) *The Emotions: Outline of a Theory*. For those interested in the relationship between phenomenology and the social sciences, see Schutz (1967).

[5] For a detailed critique of the contemporary phenomenological psychology movement in the United States, see Kockelmans (1971).

Gregory, in his chapter "The Positivist Legacy and Geography", also perpetuates the confusion between this particular position in logic and a positivistic interpretation of science when he claims that (1978, page 33):

> "Nevertheless, it should still be clear that the factual sciences were of primary importance to the logical positivists and that they were firm in their rejection of any metaphysical relativism. Indeed, they not only retained but elaborated a commitment to the nomological conception of science, that is, a concern with the inferential procedures which determined, in their most general form, *if C then E*",

where C stands for initial conditions and E is the event to be predicted. Relph makes the same substitution when he claims that "there is certainly a well-developed tradition of positivism in modern geography and this is based on assumptions that are radically different from those of phenomenology" (1970, page 198), and the assumptions he details include the assumption that positivism equals the deductive-predictive approach (1970, page 198). This confusion between positivism and Hempel's covering law of explanation is unfortunate and misleading and must be clarified. Although Hempel is, indeed, a logical empiricist, his covering law, which only accounts for the logical structure of explanation, is not inherently positivistic.

Hempel's covering law constitutes a position in logic (Hempel, 1966). One need not be, nor need not become, a positivist by accepting this covering law as an appropriate or adequate conception of explanation in an empirical science. But many geographers have confused the two, and it has been compounded by a comparable confusion between positivism and Kuhn's notion of paradigms of a normal science (Kuhn, 1970).

Hay characterizes and defines positivism in terms of Kuhn's paradigms of normal science (1979). He argues that positivism is accepted throughout the discipline as a "convenient and widely used umbrella term to describe the philosophies of normal science" (1979, page 2), and he believes further that the common target of the antipositivists is the same: "The paradigms of normal science" (1979, page 2). But such a conclusion is unwarranted. A paradigm within a normal science can be, but need not be, interpreted in a positivistic way. For example, quantum mechanics is a paradigm in contemporary physics, although it can be interpreted either in a positivist or in a nonpositivist way.

The unfortunate implication of this definition of positivism in terms of Hempel's covering law and Kuhn's paradigms is the growing belief that if one engages in empirical work using the covering law or a particular paradigm, then one must become a positivist who reduces all the sciences to physics. Although this is categorically not the case, the confusion between positivism and empirical science persists.

The only way to resolve the confusion and clarify the relationship between phenomenology and geography is to turn to philosophy itself, and there retrieve the original meanings of positivism, empirical science,

and phenomenology. Such a retrieval will lay open the possibility of
examining, quite independent of current geographical polemics, the
meaning of geography as a human science.

Positivism, scientific realism, and logical empiricism
In his "Introduction" to *The Positivist Dispute in German Sociology*,
David Frisby immediately recognizes the ambiguity, confusion, and
disagreement that surrounds the term positivism (1976, pages ix–xliv):
> "It is difficult to find a generally acceptable nominalist definition of
> the term. Positivism is not a static entity but is itself dynamic and has
> taken different forms in various historical contexts."

Discussion in contemporary human geography mirrors this same difficulty,
because it embraces the term positivism to cover three different forms—
positivism, scientific realism, and logical empiricism. In brief, positivism
refers to Comte's original formulation, whereas scientific realism and logical
empiricism characterize successively later elaborations of Comte's views.

According to Comte, man's knowledge passes through three stages of
development: the theological; the metaphysical; and the positive, or
what we refer to as the empirical or factual stage. Each scientific discipline
passes through all three stages, and the more complicated the subject
matter of a discipline, the longer it takes to achieve the third stage.
Mathematics was the first discipline to achieve the third stage completely,
whereas sociology will probably be the last. Knowledge achieved in the
third stage renders knowledge yielded by the two preceding stages both
superfluous and meaningless, since empirical knowledge achieved through
methodological rules 'guarantees' access to the absolute truth. The rules
are five-fold: (1) all knowledge must be proved by sense certainty (*le
réel*); (2) all knowledge is intersubjectively proven, in the sense that there
has to be a unity of the scientific method (*la certitude*); (3) all knowledge
must adhere to the formal construction of theories which allows for
testable hypotheses (*le précis*); (4) all knowledge must be technically
utilizable (*l'utile*); and (5) such knowledge is relative and unfinished, and
it progresses by the gradual unification and accumulative growth of
theories (*le relative*) (Gregory, 1978, pages 26–28; Frisby, 1976,
pages xi–xii). In effect, Comte claims that data exist in themselves.

Many historians and philosophers of science have disputed Comte's
theory of the stages of man's knowledge on two counts: his claim that
empirical knowledge makes theological and metaphysical knowledge
unnecessary and meaningless; and his claim that scientific data can exist
independently from the theory from which they are formulated. They
claim that his theory of the stages is arbitrary, and that even if it were
accurate it would negate the possibility of an empirical science, since it
would not allow the scientist to select what to examine or choose the
perspective from which to examine it. If the selection of subject matter
and theoretical perspective are not allowed, then the possibility of an

empirical science is negated for, contrary to Comte's view regarding positive data, no data can ever exist stripped from the meaning of the theoretical framework from which they are constituted.

Despite the fact that Comte's conception of positivism is nearly universally rejected by historians and philosophers of science, many scientists and students of science stick tenaciously to two of the ideas implied in Comte's theory. These ideas constitute scientific realism, a view which holds that empirical science teaches us the absolute and genuine truth about things and about the world in which we live, and that science should not be infected by theology or metaphysics. These two ideas accord science the status of the most privileged form of knowing, which allows science to cast the deciding vote in any conflict between a scientific statement and a theological statement.

Although a number of efforts in philosophy—such as empiricism (Hume), classical phenomenalism (Mach), rationalism (Leibniz), and critical transcendentalism (Kant)—have questioned the fundamental epistemological issues that underpin these views of the scientific realist, most scientists, implicitly or not, continue to adhere to them. Moreover, they frequently hold the logical empiricist ideal of science, an ideal which places physics as the paragon of all the sciences (physicalism), and so argues that every science, including all the human sciences, should be modeled after physics (monism). Accordingly, any empirical science should be capable of being reduced to the concepts, principles, and language of physics (reductionism). For the human sciences, this means that any term must be expressible in some observation language in which only physically relevant terms occur. Concepts referring to an individual's beliefs, feelings, attitudes, and desires must be capable of being translated into the languages of biology or physiology. When these forms of behavior are understood in terms of the concepts of physics, it follows that the human sciences do not need any language that is not used in the physical sciences themselves.

Now what passes for positivism in human geography actually encompasses all three of these belief systems about science: positivism; scientific realism; and logical empiricism. All three systems believe that science yields an absolute and unqualified truth about the world. Consequently, given a conflict between a scientific statement and a non-scientific statement, the latter would be considered as less true than the former. Logical empiricism elaborates this position when it maintains that physics stands as the ideal science that all other sciences should be reduced to. None of these positions reveal, however, that one can engage in empirical research, even if one does not believe that science yields an absolute truth about the world, or that physics constitutes an ideal science.

It is, of course, possible to do empirical research and not to accept a positivistic interpretation of science. To do so requires, however, that we penetrate to the essential meaning of an empirical science by means of ontological analysis. Ontological analysis is not concerned with the logic, methodology, growth, or politics of science. Rather, it reveals the meaning of an empirical science in terms of its *essential conditions*—conditions without which that science would not actually be empirical.

Empirical science

An empirical science yields factual knowledge that is precise, exact, and certain. The possibility of such a knowledge rests on two conditions: the theoretical framework of the science, and the attitude presupposed when using the framework—the theoretical attitude of the disinterested observer (Kant, 1965; Heidegger, 1962; and Schutz, 1962).

According to Kant in his *Critique of Pure Reason* (1965), a science is empirical because scientists lay down or posit a theoretical framework of meaning from which, and only from which, certain phenomena are taken to be legitimate subject matter for research in that particular discipline. Although Kant does not adopt the term 'theoretical framework', he captures its meaning with his term "synthetic *a priori*" (1965, pages 50ff, 70, 76, 80, 85ff, 91). The referent of both terms consists of the totality of meaning from which a scientist defines and examines facts. In the general sense, therefore, a theoretical framework constitutes the totality of meaning from which a scientist, operating in the tradition of his discipline, looks at facts. The framework is *a priori* in that it is presupposed by the scientists each time they engage in empirical work, although it is not *a priori* in the very strict Kantian sense, which would imply that it is determined quite independently of the phenomenon it studies. Rather, the concepts and methods specified by the theoretical framework are developed on the basis of principles which are intersubjectively acceptable. Moreover, they lead to hypotheses which can be tested according to a well-defined process of verification or falsification that is equally inter-subjective. Such a process of testing hypotheses yields a truth characteristic of science. But contrary to the positivist view, this conception of scientific truth is never an absolute, unqualified truth. Rather, it is an inter-subjectively certain truth bound by the original theoretical framework from which the hypotheses were formulated. It is perfectly possible to speak about truth and certitude in an empirical science. Claims are called true when they express a particular state of affairs, whereas certitude implies the degree of confidence held by the person making the claims. Since the hypotheses are formulated from the theoretical framework, which in turn is developed by means of intersubjective principles and methods, the truth generated by a science (provided it has not been falsified by relevant tests) may be taken to be an intersubjectively accessible, acceptable, and certain truth.

The scientist who engages in empirical work presupposes a distinctive way of looking at the world—the theoretical attitude of the disinterested observer (Heidegger, 1962; Schutz, 1962, page 137). As a disinterested observer, the scientist pulls back from an active engagement in the world, and in so doing he stands outside of the world, as the spectator stands outside the spectacle. The world is experienced as an object to be observed and explained, rather than as a field of domination. As a result, the scientist, *qua* scientist, does not attempt to master the world, but rather attempts to observe and explain it through the means of a particular theoretical framework. At the risk of being tangential, it is necessary to clarify this point.

One of the strongest motives for turning to and for dealing with science derives from a desire to improve the world (Habermas, 1973)[6]. The scientist often seeks to apply his scientific findings to problems in the world, and in many instances such applications have proven successful. Yet, neither the motives for, nor the applications of, science exist as part of the process of theorizing scientifically. "Scientific theorizing is one thing, dealing with science within the world of working is another" (Schutz, 1962, page 246). Our concern with the essential meaning of an empirical science limits the focus to the process of scientific theorizing, rather than to a consideration of the motives lying behind scientific inquiry, or the way scientific findings are ultimately applied.

The theoretical attitude which guides scientific theorizing allows the scientist to thematize the world from a perspective defined by the theoretical framework. From this particular and carefully chosen perspective, the scientist demarcates a certain region of the world of everyday living as his proper subject matter of observation and explanation. He thematizes a particular region of the world as his theme of research by the processes of *idealization* (making the world into ideal types), *formalization* (describing these types in terms of formal attributes), and *functionalization* (relating these formally described types to each other, usually, but not always, by means of *quantification*). When these processes are applied to the lived world, the phenomena of the lived world are reduced to idealized entities which, by definition, must be abstract in comparison to their original forms.

The adoption of the theoretical attitude characteristic of empirical science implies that the scientist experiences himself as a distanced observer who considers the world as an object of contemplation. It is within this distanced and contemplated world that ideal entities exist that are formally described and functionally related to one another. But the nature of such a theoretical attitude raises a distinct challenge to any human science. In brief, how is a science of man possible?

---

[6] The issue of the applicability of science to 'real world' political problems is considered an issue of grave concern to such critical theorists as Habermas (1971). See McCarthy (1978) for an excellent treatment of Habermas's ideas regarding science in society, and also Gadamer (1975b) and Ricoeur (1973).

The challenge of a human science

By definition, a human science must examine man in his lived world. Yet the very nature of the theoretical attitude characteristic of science ensures that a particular human science can never grasp the fullness of man's humanity nor the immediate vividness of his lived world. Science, as a second order of meaning, can grasp only a region of man's lived world, and each human science grasps its own particular area of concern—a particular region for anthropology, another for economics, a third for geography, and so on. As Schutz claims (1962, page 254):

"The theoretical thinker while remaining in the theoretical attitude cannot experience originarily[7] or grasp in immediacy the world of everyday life within which I and you, Peter and Paul, anyone and everyone have confused and ineffable perceptions, act, work, plan, worry, hope, are born, grow up, and will die—in a word, live their life as unbroken selves in their full humanity."

According to Husserl (1970), the greatest crisis confronting science today is the lack of realization of the connection between the everyday life of man and the theoretical attitudes and frameworks of science. The crisis stems from the prevailing positivist or empiricist philosophies which focus only on the scientific frameworks and fail to recognize that (Schutz, 1962, page 120):

"The basis of meaning (*Sinnfundament*) in every science is the pre-scientific lifeworld (*Lebenswelt*) which is the one and unitary lifeworld of myself, of you, and of us all. The insight into this foundational nexus can become lost in the course of the development of a science through the centuries. It must, however, be capable of being brought back into clarity, through making evident the transformation of meaning through which this lifeworld itself has undergone during the constant process of idealization and formalization."[8]

Thus, the challenge confronting every human science is to provide an interpretation which reveals the connection between the theoretical frameworks of science and the lived world.

Such an interpretation must ensure that the resulting human science is authentically scientific and human. In order to be *scientific*, a human science must operate within an *a priori* framework; but in order to be *human*, a human science must choose an appropriate framework, one

[7] Originarily is used in the sense of primordial. Primordial indicates that the noun that follows has a *privileged position*, not necessarily privileged chronologically or epistemologically. It is privileged in the sense of original meaning.

[8] As an example of how the lived world can be thematized by the geographer *qua* scientist, the reader is referred to an excellent series of four grade school geography textbooks *New Ways in Geography* (Cole and Benyon, 1974). The exercises in these four texts start the student with real life experiences, such as playing in the school yard, and show how the experiences can be thematized into objects of geographic contemplation.

which is suitable to the lived world, and which draws its leading concepts from the being of that world. Within human geography, neither positivism nor humanism meets that challenge.

Positivism ignores the lifeworld, and tends to select frameworks which are inappropriate to the way people experience their world. For example, the framework of social physics, even though it is authentically scientific, is totally irrelevant to how people actually live their lives. People do not experience themselves as particles moving through a homogeneous space. On the other hand, humanism tries to address man 'scientifically', but does so outside of all frameworks. Since the theoretical framework is the essence of an empirical science, humanism ignores science. For example, the humanistic descriptions of lived space are genuinely human descriptions, but they are not scientific. Because they are not developed from the perspective of a framework, they are not sufficiently systematized and intersubjectively verifiable to lend themselves to the type of generality necessary to be considered scientific.

In order for a human science to be both scientific and human, it must select and justify its theoretical frameworks by means of two corollary *a priori* components to the empirical component. These include a descriptive component and an interpretative component (Kockelmans, 1978).

The descriptive component must concern itself with the framework of meaning upon which all social phenomena are to be projected, so that they can become the legitimate subject matter of empirical research. The function of such a framework is to make it possible to determine (1) the legitimate subject matter of empirical research in a given social science, (2) the characteristics and relations of the social phenomena to be considered which will be relevant, (3) the research methods to be used in the explanation of these phenomena, and (4) the linguistic means which will be adequate for the precise formulation of problems and solutions, both in terms of the purely formal framework and the 'real' social world. The task of the descriptive component of a social science is to develop a framework of meaning by means of descriptive methods of the kind proposed by Husserl in his reflections on regional ontologies. We shall consider these at greater length below. On the other hand, the interpretative component of each social science is concerned to explicate the meaning which social agents in 'real' social worlds attach to their own social actions, and to subject such meaning, and the interpretation on which it rests, to a hermeneutic critique by unmasking prejudices, false beliefs, and ideologies.

The descriptive, interpretative, and empirical components are all necessary if a human science is to develop an intersubjectively valid theoretical framework that has relevance to the human phenomenon studied. Each component presupposes the theoretical attitude, yet each proceeds according to its own distinctive method. The empirical component proceeds through the method of hypothesis testing; the descriptive part develops by means of descriptive analysis; and the interpretative component employs the canons of hermeneutics. By understanding the roles of the descriptive and the interpretative components of science within a human science, we can reveal how the two orders of meaning—the scientific and the lived—can be brought together.

Descriptive science—phenomenological science

Husserl laid out the possibility and necessity of a descriptive science through his work on regional ontologies (1975, pages 9, 59, 72). The concept of regional ontologies is crucial to understanding Husserl's conception of phenomenology and its relationship to an empirical science. Unfortunately, this relationship has frequently been misunderstood.

Phenomenology has often been identified by philosophers and by scientists as an effort to reject empirical science. Merleau-Ponty is frequently quoted to this effect when he states in the *Phenomenology of Perception* (1962, page viii): "Husserl's first directive to phenomenology ... is from the start a rejection of science." But a close critique of his text shows that Merleau-Ponty did not conceive of phenomenology as a rejection of empirical research. He is, in fact, in clear agreement with Husserl.

According to Husserl, each science possesses its own domain of investigation—its region—but the scientists cannot clarify this domain for themselves using the methods and logic of empirical science. Thus, alternative methods, which are essentially descriptive, must be developed in order to reveal the essential structures of the entities under study by the discipline. For example, the region for psychology consists of the 'region of psychic phenomena'.

The totality of entities in a certain region is fixed in its 'regional categories'—categories which are typical and relevant to each region. It is in these regional categories that (Kockelmans, 1967a, page 101):

"... all the *a priori* presuppositions [are] contained under which each multiplicity of beings which are immediately given in our pre-scientific experience can be conceived and understood as belonging together in such a way that they can become the object or theme of one or the other science."

Husserl characterizes the sciences in which these categories are discovered and revealed as regional ontologies—descriptive sciences of a region of entities for a particular discipline. Thus, each empirical science possesses its own corresponding descriptive science.

The purpose of a regional ontology is to describe the domain of entities appropriate to that science. This purpose is achieved through an onto-logical description of the *a priori* theoretical framework posited by a science when it engages in empirical work. Such a description lays out precisely the origin, the meaning, and the functions of the concepts, principles, and methods of a particular framework which has been assumed before that science can establish facts, develop hypotheses, or build theory. In so doing, it achieves several purposes. It reveals the range of legitimate subject matter of the empirical science, as well as the characteristics and relations of the particular subject matter. It specifies the research methods that are to be used in constructing an explanation, and it defines the language necessary for the precise formulation of the problem statement and solution. Such a language must be appropriate both to the formal framework of meaning of the science and to the language used in the lived world by the people being studied.

A regional ontology should yield a number of fundamental theses about the phenomena studied, and about the invariable perspectives that must be embedded in the concepts of the theoretical framework. In this way, the theoretical framework presupposed in the empirical component can be universal to all the phenomena it studies, in all places, and at all times. It can be accessible to any scientist standing in that community of scientists. In addition, however, each human science must also ensure that the theoretical framework selects concepts, principles, and methods that are relevant to the human phenomena as they occur in 'real' social worlds. This requires that we introduce yet another component to science—the *interpretative* component.

Interpretative science
The descriptive component of a given human science cannot guarantee that the theoretical framework of meaning employed by the empirical component is adequate and relevant to the lived world. This can be guaranteed only by an interpretative component to science which accounts critically for the meanings held by the agents in their lived world. Such an account penetrates to the primary order of meaning—the lived world— so as to 'understand' scientifically and critically the human phenomenon which the empirical component explains. As such, the interpretative component does not provide a substitute for empirical science. Nor, more pointedly, does it constitute an exercise in empathy—an exercise which would attempt to discover the private thoughts, fears, experiences, or aspirations of isolated individuals.

The interpretative component of each social science tries to provide a scientifically and critically acceptable account of the meaning which the social agents themselves attach to their own social actions in the concrete lifeworld in which they live and act. Such an account is necessary in order to make certain that the theoretical framework of meaning provided by the descriptive component, and effectively employed in the empirical component, is indeed relevant to actual social phenomena. It is also necessary in order to eliminate prejudices, false beliefs, and ideologies to which the self-conception of social agents is usually subject. In the interpretative component of each social science, methods are used that were originally developed for the philological and historical sciences in the science 'hermeneutics'. These methods are partly interpretative and partly critical (Boeckh, 1968; Dilthey, 1962; Gadamer, 1975b; 1976; Heidegger, 1962; Ricoeur, 1965; Kockelmans, 1975; and Seebohm, 1977)[9].

Interpretation reveals how the agents construct their world—what meaning they attach to their actions and their thoughts insofar as these actions and thoughts are publicly accessible and manifest themselves in 'overt behavior'. Interpretation focuses on what is meant by the agents; it does not examine private thoughts or actions which are not subject to public scrutiny. As a component of science, such interpretation must reveal only publicly accessible shared meanings.

In its turn, critique subjects interpretation to the question "What is the precise meaning of the assumptions upon which the people's interpretations of the world rest?" Whereas the interpretation attempts to reveal what the agents themselves really meant to achieve by their social actions, the hermeneutic critique tries to examine all the prejudgements upon which this conception appears to rest. It strives to ensure that the abstract theoretical framework presupposed in empirical research, and precisely laid out in descriptive analysis, is *grounded in, relevant to*, and *systematically derived from* the concrete world in which the agents live their lives. The interpretative component provides the necessary balance to avoid the selection and superimposition of inadequate and irrelevant theoretical frameworks onto human phenomena.

Geography as a human science

A human science must be capable of connecting the two orders of meaning —the primary order of the lived world, and the secondary order of the scientific world. Yet the prevailing philosophical dogmas in human geography sever these two worlds and obscure their connection. Positivism, in its contemporary guise as logical empiricism, arbitrarily selects and superimposes the theoretical frameworks of physics onto the human world. Humanism, with its unique, and at times neoromantic interpretation of phenomenology, reveals the lived world but leaves it dangling and

[9] Although hermeneutics has undergone some conceptual reshaping, its original meaning as articulated by Boeckh in the mid-nineteenth century has gone unchanged (Seebohm, 1977, pages 181 ff). Hermeneutics is a method, and to the extent that it is developed systematically, and is justified, it is a methodology which produces an interpretation and a critique of a text or text-like phenomenon. Boeckh employed hermeneutics for the interpretation of texts, particularly the Bible. Dilthey adopted and adapted hermeneutics as a method to reveal the foundation of meaning of an empirical human science (1962). As a result, hermeneutics evolved from simply being a method for understanding texts into a method for achieving understanding in general.

Heidegger introduced the term to denote his approach to the question of Being when he speaks about the hermeneutics of *Dasein* (1962). Gadamer (1975a; 1975b; 1976) and Ricoeur (1965) elaborated this use of hermeneutics as a method for interpreting aspects of man's Being or, more simply, his modes of existence.

disconnected from the scientific. In many instances, each of these philosophical stances are premised on confused and ambiguous claims regarding positivism, empirical science, and phenomenology.

Contrary to claims made in the discipline, positivism (logical empiricism) is not synonymous with empirical science, because the sole essence of an empirical science is that it yields a precise, exact, and certain truth limited from the perspective of a defined theoretical framework. On the other hand, positivism and scientific realism accord science the status of the most privileged form of knowing which makes all other forms superfluous and meaningless and which yields are absolute truth. Logical empiricism elaborates this belief when it argues that all sciences—both physical and human—should be modeled after physics. Whereas, phenomenology involves a rejection of the realist and empiricist beliefs, it does not constitute or imply a rejection of empirical science. In fact, it is through phenomenology that we gain insight into the connection between the lived world and the world of science (Husserl, 1965; Kockelmans, 1967b).

According to Husserl, the lived world provides the foundation of meaning for the world of empirical human science. This world of empirical science explains relations in the lived world. The explanations derive from hypotheses which are formulated from the perspective of the theoretical framework. The challenge confronting a human science is to ensure that the selection of the theoretical framework—its concepts, principles, and methods—is simultaneously valid for the community of scientists and relevant to the world as lived.

The selection and the justification of the theoretical framework of an empirical human science can be achieved through two corollary *a priori* components—both of which are scientific in character, yet each of which achieve different ends and proceed through different methods. The descriptive component lays out the meanings of the concepts and the methods of the theoretical framework; in so doing it can yield some fundamental theses about the meaning of the phenomena that a science studies. This descriptive component, which Husserl refers to as a regional ontology, opens up the framework for careful scrutiny and evaluation. In so doing, it ensures the validity of the framework within the community of scientists of that discipline.

But the descriptive science circumscribes its concern to those issued raised by the scientists *as scientists*: for example, what do the scientists mean when they use particular concepts, principles, and methods; what assumptions do they operate from; and what are the implications of using them? In a sense, the descriptive science focuses on the theoretical framework only in the context of science. In a human science, such a focus on the second order of meaning is insufficient because it provides no safeguards for guaranteeing that the lived world will be meaningfully represented in the particular concepts, principles, and methods chosen.

Such a safeguard is achieved through another component to science—the interpretative.

Whereas the empirical component explains phenomena, and the descriptive component describes the theoretical framework posited by the empirical scientist, the interpretative component penetrates to the primary order of meaning—the lived world. It interprets and renders a critique of the meanings held by the agents of their world, meanings that exist independently of, and prior to, any scientific explanation of them. The interpretation and critique can be used as a means to evaluate the adequacy and relevancy of the theoretical concepts and methods to the phenomenon studied—man.

The essence of geography as a human science is its subject matter—man in his lived world. The challenge confronting geography is to provide an interpretation of science which allows geographers to choose a framework of meaning that is not only suitable to the lived world, but one that draws its leading concepts from the being of that world. Neither positivism or humanism can provide such an interpretation. But an interpretation of science which calls for all three components—the empirical, the descriptive, and the interpretative—can meet the challenge facing geography as a human science.

Acknowledgement. I would like to thank Joseph J Kockelmans for his critically insightful comments on earlier drafts of this essay. Any errors of fact or interpretation are solely my responsibility.

References
Boeckh A, 1968 *On Interpretation and Criticism* translator J P Pritchard (University of Oklahoma Press, Norman, Okla)
Buttimer A, 1974 *Values in Geography* Resource Paper, Association of American Geographers, Washington, DC
Buttimer A, 1976 "Grasping the dynamism of lifeworld" *Annals, Association of American Geographers* **66** 277-292
Cole J P, Benyon J, 1974 *New Ways in Geography* (Basil Blackwell, London)
Comte A, 1864 *Cours de Philosophie Positive* second edition (Bailliere, Paris)
Dilthey W, 1962 *Pattern and Meaning in History* Ed. H P Rickman (Harper and Row, New York)
Entrikin N, 1976 "Contemporary humanism in geography" *Annals, Association of American Geographers* **66** 615-632
Frisby D, 1976 "Introduction to the English translation" in *The Positivist Dispute in German Sociology* Eds T Adorno, H Albert, R Dahrendorf, J Habermas, H Pilot, K Popper, translators G Adey, D Frisby (Harper and Row, New York) pp ix-xliv
Gadamer H, 1975a *Truth and Method* Eds G Barden, J Cumming (Seabury Press, New York)
Gadamer H, 1975b "Hermeneutics and social science" *Cultural Hermeneutics* **2** 307-316
Gadamer G G, 1976 *Philosophical Hermeneutics* Ed. D Linge (University of California Press, Berkeley, Calif.)
Georgi A, Fischer W, Von Eckartsberg R, 1971 *Duquesne Studies in Phenomenological Psychology* (Duquesne University/Humanities Press, Pittsburgh, Pa)
Gregory D, 1978 *Ideology, Science and Human Geography* (Hutchinson, London)

Habermas J, 1971 *Knowledge and Human Interest* translator J Shapiro (Beacon Press, Boston)

Habermas J, 1973 *Theory and Practice* translator J Viertel (Beacon Press, Boston)

Hay A, 1979 "Positivism in human geography: response to critics" in *Geography and the Urban Environment* Eds R Johnston, D Herbert (John Wiley, New York) volume 2, pages 1-26

Heidegger M, 1962 *Being and Time* translators J Macquarrie, E Robinson (Harper and Row, New York)

Hempel C, 1966 *Philosophy of Natural Science* (Prentice-Hall, Englewood Cliffs, NJ)

Husserl E, 1965 "Philosophy as rigorous science" in *Edmund Husserl: Phenomenology and the Crisis of Philosophy* Ed. Q Lauer (Harper and Row, New York) pp 69-147

Husserl E, 1970 *The Crisis of European Sciences and Transcendental Phenomenology* translator D Carr (Northwestern University Press, Evanston, Ill.)

Husserl E, 1975 *Ideas: General Introduction to Pure Phenomenology* translator W B Gibson (Collier Books, New York)

Husserl E, 1977 *Phenomenological Psychology* translator J Scanlon (Martinus Nijhoff, The Hague)

Kant I, 1965 *Critique of Pure Reason* translator N K Smith (St Martin's Press, New York)

Kockelmans J, 1967a *A First Introduction to Husserl's Phenomenology* (Duquesne University Press, Pittsburgh, Pa)

Kockelmans J, 1967b *Phenomenology: The Philosophy of Edmund Husserl and Its Interpretation* (Doubleday, New York)

Kockelmans J, 1969 *The World in Science and Philosophy* (Bruce Publishing, Milwaukee)

Kockelmans J, 1971 "Phenomenological psychology in the United States: a critical analysis of the actual situation" *Journal of Phenomenological Psychology* 1 139-172

Kockelmans J, 1975 "Toward an interpretative or hermeneutic social science" *Graduate Faculty Philosophy Journal* 5 73-96

Kockelmans J, 1978 "Reflections on social theory" *Human Studies* 1 1-15

Kockelmans J, 1979 "Deskriptive oder interpretierende Phänomenologie in Schütz' Konzeption der Sozialwissenschaft" in *Alfred Schütz und die Idee des Alltags in den Sozialwissenschaften* Eds W Sprondel, R Grathoff (Ferdinand Enke, Stuttgart)

Kuhn T, 1970 *The Structure of Scientific Revolutions* (University of Chicago Press, Chicago)

Ley D, Samuels M, 1978 *Humanistic Geography* (Maaroufa Press, Chicago)

McCarthy T, 1978 *The Critical Theory of Jurgen Habermas* (MIT Press, Cambridge, Mass)

Mercer D, Powell J, 1972 "Phenomenology and related non-positivistic viewpoints in the social sciences" *Monash University Publications in Geography* (Department of Geography, Monash University, Clayton, Victoria, Australia)

Merleau-Ponty M, 1962 *Phenomenology of Perception* translator C Smith (Humanities Press, New Jersey)

Olsson G, 1970 "Logics and social engineering" *Geographical Analysis* 2 361-375

Radnitzky G, 1970 *Contemporary Schools of Metascience, I and II* (Akademieförlaget, Göteborg)

Relph E, 1970 "An inquiry into the relations between phenomenology and geography" *Canadian Geographer* 14 193-201

Ricoeur P, 1965 *History and Truth* translator C Kelby (Northwestern University Press, Evanston, Ill.)

Ricoeur P, 1973 "Habermas and Gadamer in dialogue" *Philosophy Today* 17 153-165

Rowles G, 1978 "Reflections on experiential field work" in *Humanistic Geography* Eds D Ley, M Samuels (Maaroufa Press, Chicago) pp 173-193

Sartre J P, 1948 *The Emotions: Outline of a Theory* translator B Frechtman (Philosophical Library, New York)

Schutz A, 1962 *Collected Papers, Volume I. The Problem of Social Reality* Ed. M Natanson (Martinus Nijhoff, The Hague)

Schutz A, 1967 "Phenomenology and the social sciences" in *Phenomenology: The Philosophy of Edmund Husserl and Its Interpretation* Ed. J Kockelmans (Doubleday, New York) pp 450–472

Seamon D, 1979 *A Geography of the Lifeworld: Movement, Rest and Encounter* (St Martin's Press, New York)

Seebohm T, 1977 "The problem of hermeneutics in recent Anglo-American Literature, part I" *Philosophy and Rhetoric* **10** 180–197

Tuan Y F, 1971 "Geography, phenomenology, and the study of human nature" *Canadian Geographer* **15** 181–192

Tuan Y F, 1974 *Topophilia* (Prentice-Hall, Englewood Cliffs, NJ)

Tuan Y F, 1976 "Humanistic geography" *Annals, Association of American Geographers* **66** 266–276

Tuan Y F, 1979 "Space and place: humanistic perspectives" in *Philosophy in Geography* Eds S Gale, G Olsson (Reidel, Boston) pp 387–427

Lived space and the liveability of cities

Anne Osterrieth

This essay discusses the existential relationship of man to his urban environment, and aims at establishing elements of a sound theoretical framework, applicable to the analysis of urban life and the management of urban environments. The notion of lived space is proposed as an organizing concept for the assessment of the quality of life in cities, and the individual experiences of people are taken as the center of concern.

Liveability

The spirit of this essay derives from the social legacy of the era of the 1960s. Since that time, geographers have increasingly taken on the task of urban reformers, and their field is expressed more and more as actual practice in urban planning. In the liberal ideology of the urban scientists, two streams of thought are prevalent. Some urban specialists—geographers, planners, architects, politicians—have advocated the 'liveable city'. Although their slogan "People matter most!" has popular support, the quality of life they advocate is partly class biased; leisure, culture, aestheticism, and participation tend to be the preoccupation of an emergent professional class (Ley, 1980) to which many of our urban specialists belong.

Other scholars and practitioners have interpreted the ideal of the liveable city in terms of social justice. Their analyses focus on the unequal distribution of power and resources over the city. Such concern has been incorporated in the broader question of the quality of life in cities as measured by social indicators. However, social indicators are aggregate statistics generated "as a by-product of some administrative process which at best constitutes only an indirect surrogate for the reality of human life experience which it purports to represent" (Smith, 1973, page 136). Quality of life, or liveability, is essentially a perceived state of affairs. Thus liveability should not be considered solely as a matter of urbanity or of social statistics, but should be defined in terms of the actual well-being experienced by urban residents. A liveable urban environment is one that accommodates the individual's expectations, an environment that minimizes the external constraints on people's physical and psychological well-being. Environmental satisfaction, including the physical and the social environments, is thus the clue to the liveable city. Hence the way to gain insight into the quality of life in cities is to focus on people's needs and feelings in their daily interaction with their environment[1]. The concept of lived space corresponds to this experiential approach.

[1] The issue of needs has been discussed elsewhere (Harvey, 1973; Smith, 1977). For the purpose of this text, needs refer primarily to latent demand (felt needs) and,

Lived space
A semantic analysis of lived space
The term 'lived space' is a recent addition to the geographical vocabulary, and to some of us the actual juxtaposition of the words 'lived' and 'space' may appear incongruous. Indeed, the geographer trained in a strong positivist tradition may simply consider space as something out-there, set and static, more or less abstract, with some sort of topological properties. In contrast, the word 'lived' connotes sensory involvement, emotional experience, and human action. But the combined expression 'lived space' has a distinctly intuitive appeal. The word 'space' rings in our geographic consciousness, and the word 'lived' appeals to our humanitarian conscience. The expression acts like a double Pavlovian bell; by catching our attention, it certainly serves a *connotative* role, although that is not its sole function. To understand its *denotative* content we must first review our definition of space.

Space is a word with many dictionary definitions, but all underline this fundamental idea: space is an organizing dimension allowing for coexistence, whereas time provides sequence. Space, at first, is a phenomenal construct, a fact that is supported by research in such fields as development psychology and the psychology of language. We apprehend the world through our senses. Our mobile bodily self orders our perceptual field into up and down, front and back, left and right. Based on sensory evidence, we develop spatial abilities such as recognizing places and topological relationships. Knowledge *of* space is phenomenal; but knowledge *about* space is intellectual, since the human mind can extrapolate far beyond the sensory data. The nature of the space intellectually constructed depends upon the criteria used for conceptualization, and each set of criteria reflects a particular purpose dominating thought. For example, Euclidean space is based on mathematical criteria defined to further the measuring task. Mythological or cosmological space, social space, astronomical space, and travel-time space are other examples of such conceptualizations. Thus, geographical space is only one of many possible ways of conceptualization.

A definition of lived space
Consider now *lived space*, a term originally coined to denote man's spatial experience (Bollnow, 1967). Experience is both objectifying and subjective. We dissect our urban landscape into blocks, streets, and parks, but we also construct a world centered around the home, and divide its surroundings into wilderness and tame space. Lived space is not a mere physical substratum, the locale of one's actions, but a mental construct: it is one's conception of empirical spaces as they are experienced and defined in daily situations. Most definitions arise from a social context. For example, intimacy and privacy are part of our definition of the home. Because lived space is based on personal experience, it usually centers around the home.

to a lesser degree, to potential demand ('real' needs), independently of people's expectations or lack thereof.

The home is often where our selves are, and where we are really ourselves. The home is the point of departure from which we orient ourselves, the place around which we organize the world, the refuge that we leave daily to meet others. As Buttimer (1976, page 284) puts it, "... each person is surrounded by concentric layers of lived space, from room to home, neighborhood, city, region and nation."

However, not all layers are lived with the same intensity or frequency, and differences may be particularly marked along the lines of social class, age, and sex. Nor do all layers carry the same purpose. The daily environment is where the intimate ties are woven, as the following example bears witness. A new Chicagoan said:

"Sometimes I wonder if it's just nostalgia, but I can remember the feeling of being at home when I returned to New Orleans. I'm sure it must have been part of my growing-up experience there. But it's interesting that I sometimes can feel something very similar walking on Kenmore [street], after being away a week or two on a trip ... I truly feel Kenmore is home for me now. Though our present involvement will be three years" (Terkel, 1967, page 149).

The familiar world carries most importance; the everyday environment is at the core of lived space. In the known, frequented, familiar space, we experience involvement in the world.

Lived space is a product of symbolic interaction, and is to be interpreted as such. Humans do not simply respond to environmental stimuli, but rather assign meaning to them and act on the basis of these meanings (Manis and Meltzer, 1978). The process of environmental knowing is that of making sense of the world, of imposing order and meaning upon experience. Lived space is cognitively constructed, "subjectively organized, and based especially upon an interpreted social environment" (Schellenberg, 1978, page 127). As Ley (1974) points out, life in the city does not depend only upon learning environmental clues and their spatial configurations, but also upon interpreting their social content. In a violence plagued neighborhood, the difference between a right and wrong interpretation may become a matter of life or death. Lived space is constructed on the basis of place significance, but significance is derived from use value, behavioral interpretations or expectations, and the symbolism of places (Moore, 1979). Some significances are established by social consensus enacted in mores and laws, and several levels of meaning may appear concomittantly. For example, the behavioral norm in a church is to be silent and meditative, since the activity taking place is worship. The symbolic meaning of the church is that of a sacred ground where God is present. Thus place, activity, symbolic meaning, and behavioral expectations become indissolubly linked in a situation which is socially defined. However, only in primitive societies can we map directly a social system of customs and beliefs onto a physical environment.

Contemporary societies are culturally heterogeneous, and social consensus is limited to particular groups and situations. Thus the process of attaching significance to place, through the interpretation of lived situations in relation to these places, becomes the central object of lived space research.

The tradition of French empiricism

To approach the empirical study of people's lived space, it is useful to take what we might call a 'loosely structural approach', in the sense that there is an emphasis upon semiotics, and a general acknowledgement that culture and personal mental processes determine how and what we know. A number of French geographers are primarily interested in collectively shared spatial representations. They seek to determine the referential frameworks that guide the apprehension of the environment within a cultural group. Gallais (1976), for example, has divided lived space into three interrelating planes: the ecological, the structural, and the affective [2]. The ecological dimension refers to a space constructed as a set of available resources relating to the activities undertaken by a population. The structural dimension refers to sociological space. It relates to kinship, status, allegiances, and enmities. Finally, the affective dimension encompasses emotions and symbolism attached to places, and refers to a signifying space.

Such a division of the spatial experience into three broad categories leads us to an operational definition of lived space. It tells us what to look for to understand a person's, or even a whole people's, experience of the world. For each plane the researcher will try to discover the referential framework of significances that guides the apprehension of space and organizes the lived environment. At that point, he or she will relate all the planes in an interactionist perspective, a perspective stemming from a tradition of American sociology. The physical setting (place), the participants or reference groups, and their significations (meanings, expectations) are all part of an active situation, and it is the *aggregation* of all the situations lived by a person that make up his or her lived space. The importance of relating the three planes is clear from the following example. First, a person's social network may be generated by their activities. Our Chicago resident again makes the point:

"After first moving here, we made ourselves known the way any other neighbor on the block would. We frequented the places that Kenmore frequents. We did our laundry in the laundromat around the corner, which is the real communications spot for many of the women on the block " (Terkel, 1967, page 144).

[2] Parallels can be drawn with the author's classification. Levi-Strauss discerns four schemata: the geographical, the technoeconomic, the sociological, and the cosmological (Tuan, 1972). Similarly, Butzer (1978) considers space on four levels: space as a set of available resources, space as a matter of control, space in terms of social identification, and space in terms of symbolic value.

Second, the structure of the ecological plane of one's lived space cannot be adequately understood without taking into account the person's communication networks (people, newspapers, etc). Such networks are the particular person's main sources of knowledge about potential resources. In fact, the very definition of resource may depend on such reference groups. The social networks and the definitions of resources are both crucial for job hunting, and for making residential choices (Duncan and Duncan, 1976). Third, work and consumption patterns may be so linked to group allegiance that the structure of the ecological plane can be understood only in terms of the sociological and the affective dimensions. For example, one may not wish to exercise the option of leaving an underpaid and tedious job because it would entail losing one's close circle of work friends. Last, activities geared to the satisfaction of pleasure create an affective bond with the place where they occur. The sights one likes to contemplate, the paths along which one jogs, the store where one gets candy, the house of a loved one, all these become significant spaces in terms of use and affectivity. We may conclude, then, that it is the relationships between the three planes that structure the reality lived by people, and that the lack of, or a negative relationship to, such planes is a source of alienation.

The operational definition of lived space into three dimensions allows for the comparison of different group experiences as we compare their structures of reality. Such research would be particularly helpful as we try to understand land-use conflicts and problems in cooperative development. The works of Johnson (1977) in Mexico and of Schwartz (1976) in the Ivory Coast provide good examples. Johnson studies two conflicting groups, villagers and government technicians, and evaluates their differences in terms of resource assessment. A semantic analysis of their respective environmental terminology shows the technician's blindness to local botanical richness and local ecological practice. Schwartz, on the other hand, explains how efforts in cooperative development in Guéré villages have failed because of the administrations misperception of Guéré social space. The developers consider the village as an integrated sociospatial unit on which to base development efforts, while "in fact the Guéré village has become a heterogeneous agglomeration of patriclan fragments" (Schwartz, 1976, page 21). Since the limits of the partriclan are the effective limits of interpersonal transactions, they partition the village with invisible boundaries, and extend the patriclan's lived space to several villages.

Unfortunately, most research on lived space does not live up to the standards set by Gallais (1976) and Schwartz (1976). In French geography, a common mistake has been to objectify lived space into a mappable environment. For example, de Golbéry (1976) investigates the partitioning of geographic space into social and familial areas in India and yet labels his work a study of lived space. In much of the French literature, as

Chevalier (1974) points out, there is confusion between 'lived space' (*espace vécu*) and 'life space' or frequented space (*espace de vie*). Life space is the areal extent of economic activities and social relations. It corresponds to the daily contact space or activity space in much of the American social science literature. Lived space, on the other hand, is a space image, a representation loaded with values.

However, lived space is not to be confused with mental maps, as some researchers are tempted to do (for example, Metton, 1974; Metton and Bertrand, 1974). Lived space is more than factual knowledge of the environment. Lived space is experiential knowledge encompassing emotion, meaning, and expectations towards places. Lived space is not to be seen from the point of view of visual memory, but as the product of ongoing symbolic interaction.

Other common mistakes found in the French literature are the tendency to reduce lived space to one of its dimensions (ecological, sociological, affective), and to restrict lived space to familiar and secure space (for example, Metton and Bertrand, 1974). In the latter case, the affective dimension is limited to positive affect towards place. It ignores the possibility that alienation, fear, drabness, and so on may be important components of environmental experience. Positive and negative sides should both be considered. Positive affect demonstrates the merging of human identity with the surrounding environment, which makes certain places so significant to their residents. Negative affect expresses alienation from place, creating a feeling of placelessness which can ultimately destroy a sense of self, hope, and worth.

Functionalism

In the American literature, man's spatial experience has been interpreted both in functional and in phenomenological terms. Functionalism, which prevails in geography, postulates man as a socioeconomic being that "behaves predictably within the norms of society" (Tuan, 1972, page 330). Behavior and processes can be observed and explained as functional relations, and such functions are seen almost entirely in economic and biological terms. In the economic sense, man's relation to his environment is one of place utility; "... the value we attach to any given place can be defined as a place utility. It is the nature and spatial extent of all our place utilities that comprise our action spaces" (Jakle et al, 1976, page 92). Thus action space is the economic equivalent of lived space. A variant of this functional framework is found in the ethological approach, where the needs for food, shelter, and reproduction are seen as satisfied via territorial behavior (a universal animal trait). Humans may display their territoriality in discrete ways. The home serves the need for rest and reproduction, while the home base, a larger area which may be actively defended by its occupants, serves as their zone of security. Beyond this appropriated space lies the home range, an area where economic and social activities are still undertaken, but where the individual demonstrates little territorial behavior (Porteous, 1977). Home base and home range correspond to the concept of lived space in the ethological perspective.

Yet these functional interpretations are essentially different from our notion of lived space. They narrow down man–environment relationships to a single mediating factor (be it economic or territorial), which does not do justice to the multiplicity of motives of human behavior. Once the functional principle is set, the derivative theory is ill-equipped to explain what might be labelled *a*functional and *dys*functional behavior. The problem stems from the fact that functionalism does not give human consciousness a central role. It does not acknowledge the importance of communication networks and group belonging which create shared perspectives between individuals. In other words, it does not deal adequately either with the individual or with the social constructions of reality.

Phenomenology

As we saw earlier, human spatial experience can be interpreted either in functional or in phenomenological terms. It is in the literature of the latter perspective that one finds reference to lived space. Phenomenology is essentially interested in human consciousness and human experience, and from this philosophical perspective some thing *is* as it appears, as it actually takes meaning in our consciousness. Thus, any thing or event becomes interesting and pertinent only to the extent to which it has a particular significance for man. Entrikin translates this idea into spatial terms. He says: "... the extension of man's spatial experience is the experience of involvement in the world" (Entrikin, 1976, page 624). Involvement encompasses emotion, intent (or purpose), and meaning. "The distance of existential space is thus the measure of obstacles faced by an individual in achieving his goals" (Entrikin, 1976, page 624). Such a concept of distance is as much emotional as functional, as the following examples demonstrate. For people who are foes, antagonism and distrust create a feeling of distance between the individuals involved, and the hostile tension also engenders a sense of crowding and narrowness of the place setting. Love, on the other hand, creates a feeling of closeness and intimacy, and yet at the same time "generates space, breath and freedom" for the lover (Rilke, in Bollnow, 1967, page 186). Thus human emotion is an integral part of a person's apprehension of the world.

The strength of the phenomenological approach resides in its explicit attention to emotion, purpose, and meaning in spatial experience. Interaction with the environment is significantly affective, and this process transforms the consciousness of economic and social relations. Seen in this light, the essence of lived space resides in the interaction with the daily environment, for the familiar world carries the most personal significance. However, the phenomenological approach to lived space has its own problems. In describing human spatial experience the phenomeno-logist too easily transposes his own subjectivity on other people's experience. Relph is a good case in point: in his eyes, anything new and mass produced—especially in the realm of housing—becomes "kitsch" and inauthentic, and it erodes "... existential insideness by destroying the bases for identity with places" (Relph, 1976, page 58). Relph errs on two counts: he is guilty of environmental determinism, and he is guilty of imputing his own historical sentimentalism to the rest of society.

Subjectivism is the major reef upon which phenomenological writing may founder. Vagueness and abstraction are the other problems, as definitions tend to become impressionistic. Lived space is said to be "... a lived horizon along which things are perceived and valued" (Schrag, 1969, page 55). In Relph's terms, "... lived space is the inner structure of space as it appears to us in our concrete experiences of the world as members of a cultural group ... . It is the space in which 'human intention describes itself on earth" (Relph, 1976, page 12). Such definitions carry little conceptual value, and no operational value—a problem made worse by the scarcity of empirical studies in phenomeno-logical geography. Phenomenology, then, serves as a source of philosophical inspiration at best, and we must turn to an interactionist model for operational guidelines.

In brief, the nature of lived space can be described as follows: it is one's conception of empirical spaces as they are experienced and defined by one's involvement in daily situations. Because lived space is based on personal experience and significance, it centers around the home, and corresponds to the familiar world. Significance is derived from use value, behavioral characteristics, and the affect and symbolism attributed to places. Lived space consists of three interrelating planes: the ecological, the sociological, and the affective. It is the relationships between these three planes that structure reality.

Liveability as definition of the situation
It only remains to add that if lived space is best seen in situational terms, the affective dimension, given its subjective evaluative nature, corresponds to the individual's definition of the situation. It is in the relationship of the affective dimension with the ecological and the structural that the degree of environmental satisfaction is expressed. Buttimer (1972), in her study of Glasgow, has shown how residential satisfaction was embodied in the feeling of 'at homeness', and how this feeling is related to proximity of kin and friends for certain people, whereas for others it requires interaction generated around services. Thus the affective dimension, with its relationship to the ecological and the structural, expresses the degree of environmental satisfaction, so the degree of satisfaction is a definitional component of situation (Deseran, 1978).

To bring this point a step further, it is necessary to introduce Michelson's (1970) concept of experiential congruence, which is attained when the environment actually accommodates the characteristics and behavior of people. To determine the degree of experiential congruence is to discover how well the social and physical environment fits a person's or group's activities and expectations in life. To do this, the ethnographic method may be the most appropriate. A combination of participant observation, intensive interviewing, and unobtrusive measures should allow the researcher to reconstruct people's time geographies, so disclosing who did what, where, when, and with whom. Such a combination of approaches also discloses people's assessment of situations in terms of their own emotions, purposes, expectations, and, of course, in terms of their personal satisfaction. Discrepancies between possible environmental opportunities and people's actual actions and yearnings could thus be noted. We would then have a more faithful picture of the quality of life in cities, and might have a clearer grasp of what needed to be changed to improve the match between an environment and its inhabitants.

Research focusing on people's evaluation of reality inevitably faces the issue of false consciousness. People may misconstrue the causes of their present state of affairs; they also may be ignorant of the socially unjust situations they live in, and therefore have low levels of expectations; or they may have unrealistic or detrimentally deviant expectations. The interviewing process is nevertheless valuable because it captures the personal interpretation of reality—the only one that is relevant to the interviewee—and because it may show the degree to which personal interpretation is a coping mechanism to deal with an unfavorable reality. Participant observation and unobtrusive measures would supplement the information needed to discover the structural relations in a person's lived space. Interviews with local specialists in service delivery may also provide a reliable source of information on real life conditions and real needs.

A focus on lived space serves essentially a diagnostic purpose, in that it represents an assessment of the quality of life experienced by urban residents. However, its findings have no binding value; ultimately the prescription for change will be based upon the diagnosis written by the policymaker. Also, the concept of lived space has limited normative value, since it does not tell us the degree to which environmental congruence is attainable. Nor does it offer direct normative precepts to deal with scarcity and conflict, whose resolution falls under the domain of social justice (Harvey, 1973). Justice is the principle of 'to each his share'. Its function resides in the allocation of resources and burdens (Smith, 1977), and its application is limited to the sustaining sphere—that is, resources such as jobs, housing, and services administered by the dominant society. It ignores the nurturing system on which psychological well-being is dependent.

The normative value of the lived-space concept is actually an indirect one. First, by encompassing both material and emotional spheres of life, it focuses on well-being, a principle of higher order than justice. Second, by the high quality of the diagnosis it offers, it ultimately has an impact on the remedies to be chosen.

We have seen how lived space is a concept of theoretical and empirical relevance. Based on an interactionist perspective, it paints a realistic picture of man–environment relationships, and provides the conceptual background often lacking in attitudinal surveys and urban ethnographies. It may provide the operational guidelines necessary to assess the liveability of cities, and these assessments can subsequently be used to bring about change. Research on lived space is ultimately intended to create a truly humanized landscape.

References

Bollnow O F, 1967 "Lived space" in *Readings in Existential Phenomenology* Eds N Lawrence, D O'Connor (Prentice-Hall, Englewood Cliffs, NJ) pp 178-186

Buttimer A, 1972 "Social space and the planning of residential areas" *Environment and Behavior* **4** 279-318

Buttimer A, 1976 "Grasping the dynamism of the lifeworld" *Annals, Association of American Geographers* **66** 277-292

Butzer K, 1978 *Dimension in Human Geography* (University of Chicago Press, Chicago)

Chevalier J, 1974 "Espace vécu et espace matrimonial en Inde" *L'Espace Géographique* **3** 68

Deseran F A, 1978 "Community satisfaction as definition of the situation: some conceptual issues" *Rural Sociology* **43** 235-249

Duncan J, Duncan N, 1976 "Housing as presentation of the self and the structure of social networks" in *Environmental Knowing* Eds G T Moore, R Golledge (Dowden, Hutchinson and Ross, Stroudsburg, Pa) pp 247-253

Entrikin N, 1976 "Contemporary humanism in geography" *Annals, Association of American Geographers* **66** 615-632

Gallais J, 1976 "De quelques aspects de l'espace vécu dans les civilisations du monde tropical" *L'Espace Géographique* **5** 5-10

de Golbéry L, 1976 "Espace vécu et espace matrimonial en Inde" *L'Espace Géographique* **5** 11-19

Harvey D, 1973 *Social Justice and the City* (Edward Arnold, London)

Jakle J A, Brunn S, Roseman C, 1976 *Human Spatial Behavior* (Duxbury Press, North Scituate, Mass)

Johnson K, 1977 *'Do as the Land Bids': Otomi Resource Use on the Eve of Irrigation* Ph D thesis, Clark University, Worcester, Mass

Ley D, 1974 *The Black Inner City as Frontier Outpost* Monograph 7, Association of American Geographers, Washington, DC

Ley D, 1980 "Liberal ideology and the post industrial city" *Annals, Association of American Geographers* **70** 238-258

Manis J G, Meltzer B N, 1978 *Symbolic Interaction, A Reader in Social Psychology* third edition, revised (Allyn and Bacon, Boston, Mass)

Metton A, 1974 "L'espace perçu: diversité des approches" *L'Espace Géographique* **3** 228-230

Metton A, Bertrand M J, 1974 "Les espaces vécus dans une grande agglomération" *L'Espace Géographique* **3** 137-146

Michelson W, 1970 *Man and His Urban Environment* (Addison-Wesley, Reading, Mass)

Moore G T, 1979 "Knowing about environmental knowing: the current state of theory and research on environmental cognition" *Environment and Behavior* **11** 33-70

Porteous J D, 1977 *Environment and Behavior* (Addison-Wesley, Reading, Mass)

Relph R M, 1976 *Place and Placelessness* (Pion, London)

Schellenberg J A, 1978 *Masters of Social Psychology* (Oxford University Press, New York)

Schrag C O, 1969 *Experience and Being* (Northwestern University Press, Evanston, Ill.)

Schwartz A, 1976 "Espace vécu, espace villageois et développement dans la forêt ouest-ivoirienne, le cas des Guérés" *L'Espace Géographique* **5** 21-26

Smith D M, 1973 *Geography of Social Well-Being* (McGraw-Hill, New York)

Smith D M, 1977 *Human Geography, A Welfare Approach* (Edward Arnold, London)

Terkel S, 1967 *Division Street: America* (Avon Books, New York)

Tuan Y F, 1972 "Structuralism, existentialism and environmental perception" *Environment and Behavior* **4** 319-331

Perspectives on structures and relations

Is it necessary to choose? Some technical, hermeneutic, and emancipatory thoughts on inquiry

Peter Gould

> ... the way to enlarge the settled country has not been by keeping within it, but by making voyages of discovery ...
>
> Augustus de Morgan, *The Differential and Integral Calculus*

> After the journey we can fill the map
> We shall not need; that map can only show
> The journey that we need no longer go ...
>
> Elizabeth Jennings, 'Map-Makers', in *A Way of Looking*

Postscript as preface

Sometimes, after you have started writing, an essay begins to take on a life of its own, and you wonder how much control you actually had over it. That is what happened here, as I put together a mixture of explanation and speculation. Much of it is written at a personal level, partly for egotistical reasons I am sure, but partly because I sense that a general malaise accumulates from individual retrospection and dissatisfaction. As Helen Couclelis notes in this volume, intellectual disciplines turn to philosophy when they are in trouble. I think this is true; but I cannot help feeling that many of our troubles today would not have arisen in such confrontational terms if we had had an accepted tradition of open philosophical reflection as part of geographical thinking, teaching, and inquiry. I hope in future that such appraisal will not be confined to a small group, but will become a continuing and accepted activity for all students.

A winter of discontent

I am fundamentally interested in the act of inquiry and in the problems of communicating inquiry to others, particularly when the questions have human beings centerstage. To these ends of my professional life I am essentially a methodologist, prepared to beg, borrow, and steal any thing, any technique, any perspective that appears to further the inquisitive act. I prize, like the morning dew on a spider's web, those occasions of *eureka* (Beer, 1976), the sudden moments of fresh knowing when understanding breaks through, moments when you say, laughing, "How neat! I never thought about it that way before". Sometimes they come without warning, as patience is finally rewarded, as things once disparate suddenly juxtapose, as burrs of ideas stick and cling after chance encounters. They come from Marc Kac (1968) and his flea-ridden dogs; they come from Roland Barthes (1957; 1977) and his way of looking; from René Thom

(1975) and his knife edges of collapse; from Carl Jung (1961) and his childhood view of Pan ... and many others. Occasionally you can feel such moments coming, you know they are there, just around the corner. Algebra is sometimes like that, perhaps because it demonstrates how things that seem different on the surface are really the same underneath. There is a quickening as one form is seen to transform to another.

And then the orgasmic moment quietens, the sun that made the sparkle dries the dew, and you absorb the experiences, taking them for granted as they become a part of yourself. The moment of recognition fades, although later you may try to recall the difficulties, the circumstances of that chance moment of sudden seeing, and you try to use your faded eurekas to help others see the same. We call it teaching.

These are the moments that come from acts of inquiry, that can, with goodwill and prior preparation, be conveyed. Other moments of understanding appear without inquiry, and cannot, in their essence, be transferred. You do not teach the magic of Judith Jameson dancing for Maurice Béjart; you cannot convey to another the hope of that fourth movement in Nielsen's Third; nor the hot dust and small of Kim's world; nor the summer lassitude of Marcel's Balbec; nor the mythological enchantment of Flecker's "Old Ships". You cannot give to another the inner meaning, the erotic abandonment, of Klimt's *Virgin*, nor the hollow sadness of Schiele's *Family*. They are simply there. One does not inquire after them, but only responds with what one has.

A strange way to start an essay of philosophical reflection on geography, but it may be important to think about those things that can be shared, and those things that are contained within themselves, things that invite response, not inquiry. Even the languages are different: there seems to be a cover set for inquiry, and a cover set for expression, with all the overlap that such terms imply. In order to inquire, we seem to be confined to words, graphics, and algebras; these are the languages of our investigations, the tools with which we shape the knowing of our world (Gould, 1976; 1977a). But we all understand there are others: apart from our precious words, we have dance, film, painting, bodily movement, gesture, and perhaps even music, to *express* ourselves, to push out from us those things inviting response from others. In contrast, inquiry seems to ingest and devour the world. It is an aggressive, rapacious act stemming from our Greek legacy, an act driven by a curiosity that feeds upon itself. There seems to be no stopping it, even unto death (Steiner, 1978).

What do we mean by inquiry? Surely it is the ability to record observations, and so structure them that they tell a satisfying story (Gould, 1979a). I enjoy inquiry, the act of describing well so that things not seen, or only suspected, become clearer and more convincing. I enjoy struggling for the assent of the reader, even if the assent is reluctantly and critically given. As Plato noted long ago in *Timaeus* (page 29), what we would perhaps call science today is essentially telling a good story—

a story told convincingly, with such evidence and argument that a particular description is acknowledged to stand—at least for the moment[1].

I inquire for pleasure, although on occasion I feel obliged to put on the mask of serious dedication, and with grave mien and serious voice talk about being useful, about service to humankind. Such nobility of purpose is expected, and appears to make others more comfortable. But I inquire after those things that interest *me*, and if the descriptions turn out to be useful to others so much the better. I only hope financial constraints never bind so tightly, or public research funding becomes so immediately directive, that I am forced to make inquiries into things that do not interest me. I can think of nothing so self-destroying. And I make no distinction between geographic and other inquiry, and will not constrain my pleasure-seeking to topics inside the nineteenth-century box—no matter how eclectic and capacious it seems to be. Geography, like many of the other academic partitions, seems to be a good place to grow from, but a dull place to reside in. Many interesting areas of human experience lie outside of geography, and I see so little evidence of imaginative inquiry, technical competence, or reflective and informing introspection among the specialized practitioners of adjacent fields that I become less afraid of making a fool of myself. It all seems to be wide open.

But inquiry itself can devour the inquirer as it makes continuing and fresh demands. About four or five years ago the momentum that had sustained me for the past twenty years ran out. Perspectives that once seemed to liberate thought, now closed in and constrained. Mainlining game theory, entropy maximization, geometric and goal programming no longer produced the same highs. They had their eureka moments when they were fresh and new, and I shall always be grateful for the intellectual challenges they posed. The perspectives are still there, absorbed and informing when they are needed. But the data requirements were often astringent, and I had the feeling that the world was being forced into frameworks that made increasingly unacceptable distortions.

Not that I was unaware of the arguments in praise of simplifying complexity. I knew them all by heart—after all, I had taught them for nearly two decades. I had ready-made liturgies and convincing defenses for numerical taxonomy, for multivariate analysis, for system simplification, for black and gray boxes, and for optimization models in six different heuristic and proven varieties. I could defend the first deterministic steps of Ashby (1956), argue convincingly for the courage of Forrester (1971) and his counterintuitive effects, tell Ackoff's (1978) anecdotes like an old *raconteur*, wax enthusiastic about the gravity model buried in Wilson (1974), point to the ignorance times and epidemics of uncertainty in Linhart (1973), and explain why Monte Carlo simulation

---

[1] In keeping with the custom of classical scholarship, page reference to Plato's writing is based upon the 1578 edition of Stephanus; see Lee (1979, page 28).

was the *only* way to handle multichannel queuing in networks and diffusion in heterogeneous spaces (Ford and Fulkerson, 1962; Hägerstrand, 1953).

And underneath it all I was bored stiff.

Perhaps this happens sometimes, but the question then is 'what next'? There was little in traditional approaches to Geography that made the slightest intellectual demands. Occasionally a paper posed an imaginative question, and caught one's eye with a thoughtful, unexpected perspective useful for elementary teaching. But the exercises in seventeenth- and early nineteenth-century mathematics seemed to grow in the new journals, while the old ones still published the same stuff that had turned off so many of us in the late 'fifties (Gould, 1979b). There was little to sustain one there. In other fields of social science the same problems were apparent—even if the practitioners seemed unaware of them. A strong, emotional, and largely ignorant reaction to things 'quantitative' developed (Marchand, 1974), but little in the way of challenge appeared to replace the inadequacies and naiveties that came from the first flush of statistical enthusiasm. It was cold, and the ground was as hard as iron.

Spring thaw

Four strands were eventually laid together to form a rope strong enough to pull me out. First, Structuralism, in all its many faces, took me far from familiar ground, and yet showed me how other traditions of inquiry were groping towards the same questions—questions of finding the deeper patterns, the ones that lay underneath to produce different manifestations at the surface. Part of this perspective was in the tradition of good science, but the drive for such pattern-seeking came from then unexpected quarters: from literary criticism (Barthes, 1963; Steiner, 1970), from anthropology (Leach, 1969), from psychoanalysis (Eriksson, 1958; Jung, 1964), and from linguistics (Chomsky, 1968). These acknowledgements have been made elsewhere [2].

The second strand was philosophy, systematically followed according to my own, quite particular system; a system incomprehensible, and perhaps even derisible, to a professional philosopher. The continuing adventure confirms my intuition that it matters not where one starts, although an undergraduate rememberance of Plato's shadows helps—hardly surprising in the circumstances. Wittgenstein was crucial, once I understood his deeper and wholly ethical purpose [3], and when I made the reassuring

[2] Two years of teaching an interdisciplinary seminar with Alan Knight on Structuralism, entitled *Mapping the Two Cultures*, did me no harm at all, although what the poor students learnt I have no idea. I have tried to acknowledge these years before (Gould, 1979a).

[3] Alan Janik's and Stephen Toulmin's *Wittgenstein's Vienna* (1973) was a work that opened up a treasure house of intellectual connections, a pleasure dome later explored in a two-year faculty seminar at Penn State ranging over physics, music, literature, art, journalism, philosophy, poetry, theatre, photography, history ... .

discovery that I was sharing his concern for language, words, and meanings with others (Olsson, 1980; Wilson, 1976). But the foremost question of how to inquire remained, and much was skimmed and judged irrelevant to this central task. Most of it was tedious, if not tendentious; interesting, some of it, yes; but not relevant, not directed, to the act of inquiry. Sometimes it seemed that too many philosophers of science had never actually done any science, and too many phenomenologists seemed out of touch with actual phenomena. And then came Habermas (1971) casting a wider net of concern, yet catching within it a collection of problems that seemed relevant and connected. The light at the top of the well became a bit brighter, and it did no harm that connections were being made by others to the same perspectives of inquiry (Gregory, 1978; Melville, 1976).

The third strand, and one I want to follow at some length later, was the language of structure, or Q-analysis, developed by the mathematician Ronald Atkin (1974a). It has been a wrenching, decisive experience, one that has changed the way I look at the world, and so inquire about it. Equally important for one who professes to profess, it has altered what I can teach with conviction, as opposed to what I can purvey as a foil. Some mistake this algebraic topological language for 'another technique': it is not. Rather it is a way of looking, an entire methodological perspective that casts a devastating light upon the inadequacies of conventional approaches. For this reason it is not always popular with those who have made large intellectual, not to say emotional and psychic, investments in the bulging bag of multivariate techniques, with all their factor analytic, taxonomic, and other variations (Gould, 1981a). It also provides a parallel perspective upon the important idea that sometimes things are subject to catastrophic collapse (Thom, 1975; Poston and Stewart, 1978), and it will not surprise me if this general way of looking helps us to think about the emergence of order from disorder (Nicholis and Prigogine, 1977).

The final strand was constructed from acts of investigation themselves, from getting involved in real problems that demanded that you get the mud of facts and observations upon your shiny intellectual boots. For me, inquiring *with*, as well as inquiring about, Q-analysis has been a crucial, but by no means unique and individual, step. John von Neumann has commented how even in mathematics one must eventually return to the cool oasis of empirical fact after a long sojourn in the painted desert of abstraction. As beings in the world, we must inquire about inquiry, and reflect upon reflection. At the same time, we must remember that a return to the phenomena themselves is eschewed only at the price of increasing irrelevance. And so investigations have ranged over international television (Gould and Johnson, 1980a; 1980b), the taxonomy of foraminifera (Gould, 1981a), the structure of team games (Gould and Gatrell, 1980), preinvestment studies in agriculture (Gaspar and Gould,

1981), speculations in gender research (Gould, 1979c), and introductory
reviews (Gould, 1979e; 1980). In each case, the structural perspectives
helped me to see further, and marshalled evidence for the seeing—quite
apart from providing the pedagogic privilege of working alongside those
who had pioneered the way[4].

Two strands—those of philosophy, and of *Q*-analysis or polyhedral
dynamics (Atkin and Casti, 1977)—were particularly important, because
they seemed to reinforce in a symbiotic fashion the pertinence of each to
the basic questions of inquiry. Without explicitly focusing upon a major
theme of critical theory, namely the intellectual consequences of the
historical and social matrix, let us recall the perspectives of inquiry of
Jurgen Habermas (1971; 1979), noting immediately that they are
separated by him only for narrative convenience. Three are recognized:
the *technical*, the *hermeneutic*, and the *emancipatory*. The first is
associated with the empirical–analytical sciences, and is characterized
essentially by the formulation of statements about the covariation of
events. It is a tradition of inquiry typical of the physical sciences since
the seventeenth century, areas of investigation with a deep concern to
predict, and, therefore, to control. I think it is legitimate to consider
within this perspective everything from the ethnoscientific knowledge of
traditional agricultural communities, to the most technically advanced
space, atomic, and biomolecular knowledge today. All practitioners, from
the Indian farmer sensing the onset of the Monsoon, to the theoretical
physicist working at the strange frontiers of quantum theory, try to
formulate an answer to the question, 'What happens if ...'?

It is a tradition of inquiry that today generates deeply emotional
responses in many, reactions that are essentially negative and highly
fearful that the predictive perspective will be used in the realm of human
behavior to manipulate and control. Since every culture, and particularly
every religion, has always attempted to do precisely the same thing, we
actually have nothing new here. Exhaustive comment is available on the
parallels between magic, science, and religion to control the human and
physical environments, but science—ethno- or modern—frequently appears
to work better than potions or prayers, and so perhaps is feared more.
We should be acutely aware, however, that the perspective of technical
inquiry is not dismissed by Habermas himself: on the contrary, the
methodological rigor of postpositivist inquiry has been called "the
irreversible achievement of modern science" (Habermas, 1973).

The second perspective, the hermeneutic, is founded upon the act of
interpretation, upon the ascription of meaning to a text, where the term

[4] I acknowledge with deep thanks the conversations with Ronald Atkin and Jeffrey
Johnson, and the restorative contribution of Messers Adnams, so conducive to
clarifying thought.

'text' must itself be defined and enlarged to include a body of 'factual' material. Such material is simply there, a corpus of observations and statements: but what does it *mean*? Meaning implies an impression of interpretation upon such a corpus, an interpretation formed both by the particular approach of the interpreter, as well as by the particular sets of statements somehow selected for interpretation. Controversy arises because different traditions, values, and perspectives inform the approach, and this is as true of the once-radical reinterpretation of Racine by Barthes (1963), as it is of Lorentz's refusal to accept the new perspective of Einstein (Whittaker, 1960). But we must go slowly here, and not be too glib. Problems of meaning, and conflicts of interpretation, clearly do arise in the history of inquiry into the physical world, but there appears to be considerable agreement about the way to clarify the conflict, and decide which of competing interpretations is to be preferred. Both the experimentalist with the ability to design a critical test (Morrison, 1957)[5], as well as the patient observer and gatherer of new and relevant facts (Sayre, 1975), are honored here for their essential decision-provoking capacity.

But in the world of human consciousness and inquiry, as opposed to the simpler physical realm, conflicts of meaning and interpretation may be less easily settled. The latent ideologies, the unexamined discourses, are deeper and more informing of the hermeneutic task. Interpretation of text sets up an always-contingent structure, but the striving "toward the attainment of possible concensus among actors [even] in the framework of a self-understanding derived from tradition" (Habermas, 1971), is founded much more upon verbal argument, "rational persuasion", and polemic. In brief, the hermeneutic tradition seeks the assent of the reader, not through the critical experiment that seldom can be controlled or even designed, but through argument embedded in ideology—where 'ideology' itself has the entirely neutral, nonpolitical meaning of 'unexamined discourse' (Gregory, 1978). Interpretations informed by different ideologies stand parallel, and the moment one chooses or prefers, so at the same moment one accepts, perhaps just for the moment, the unexamined discourse lying behind the given meaning.

And it is here that the third and crucial perspective upon inquiry arises: the critique of ideology, and the examination of the inquiring act itself, define the emancipatory perspective. Such a backing off from the immediate technical and hermeneutic tasks, such a from-the-outside-looking-in perspective, appears essential if deeper insight and understanding are to be achieved. Not only is the hermeneutic task itself informed, as the meaning of meaning is clarified by embedding a particular interpretative perspective in the underlying, perhaps latent, ideology, but the very

---

[5] Morrison's essay refers to the experimental genius of Chien Shiung Wu to design an experiment confirming the conjectures of Lee and Yang about parity.

assumptions underlying technical acts of choice and observation may be
teased apart. And by such a teasing, such a disentangling, we come to
question the appropriateness of particular traditions of inquiry for the
human world.

We see how we have mapped human inquiry onto the paradigm of the
physical sciences, searching for laws of human behavior, or at least
believing in the more qualified phrases about 'law-like regularities'. And
here we are in an intellectual arena of terrible and unresolved tensions,
for even as we acknowledge the accuracy of many descriptions of human
mass behavior cast in terms of social physics, we recall the ability of
human consciousness to break such regularities. "The critique of ideology
... [takes] into account that information about law-like connections sets
off a process of reflection in the consciousness of those whom the laws
are about" (Habermas, 1971, page 310). After this, how can we ever
believe in the search for laws of human behavior again?

But now the question is what is left; what is left of imaginative, secure,
and useful inquiry into the human world, once the falsity of the physical
analogy is exposed? How does one proceed in an intellectually and
methodologically rigorous way, to make knowledge replicable, to make it
secure against critical scrutiny, to make that 'achievement of modern
science' truly 'irreversible'? Do we abandon the technical perspective in
favor of the hermeneutic? And, if dissatisfied with the multiplicity of
meaning, do we retreat still further, out of the arena with its blood and
sand, to find refuge in the grandstand, a place above the common people
from which we can wave our scented handkerchiefs of condescending
critique? After all, being more intelligent, aware, and sensitive than those
who sleep in darkness ... surely we have the privilege of choosing such a
life of explaining to others how terribly inadequate their own inquiries
are?

But this is a coward's way, and in time the spirit softens from such
dalliance. More important, the assumption that we must choose one
perspective over another can only be false. So I want to argue that we
do not have to choose, that we do not have to abandon the technical in
favour of the hermeneutic, and favour the emancipatory over both.
Indeed, our inquiries (I am tempted to say our lives), are diminished by
such categorical choices. We can use, *must* use, all perspectives, all
traditions of inquiry simultaneously in troikalike parallel to enrich our
understanding of that most complex phenomenon—ourselves. And I
believe there is a way, a methodological viewpoint which not only allows,
but requires, their parallel use to obtain the rigour of the technical
perspective, to achieve that meaning provided by the interpretative stance
of the hermeneutic, and to gain the self-reflective questioning of the
emancipatory. It is a young methodology, an approach just beginning,
but one that has the potential to open up an avenue of inquiry in the
human realm not really perceived before. In brief, I would like to

embed Q-analysis in the contextual setting of these perspectives of inquiry, and try to explicate, to unfold, the steps of definition, interpretation, and self-reflection that run parallel and deep.

Atkin's Q-analysis

Q-analysis employs a language of structure, a language using standard notations of algebraic topology to help us describe, think about, and interpret structural matters in a clear, unambiguous, and, above all, operational way. It is not a *technique* in any of the common senses of that word, but a complete methodological perspective, a way of looking, a way of approaching the task of description. Despite the apparent abstraction of its notation, it is fundamentally rooted in the actual and empirical world of hard data and observation. In brief, and quite paradoxically, this high-level and abstract *qualitative* language does not allow one the luxury of excessive notational speculation, but forces the user to consider the concrete reality of actual data sets. Alternatively, it points up with distressing, and often embarrassing, clarity the fact that data appropriate for more conventional techniques are not available (and probably would not be available until the year 2030 if we start collecting today), and it imposes a discipline that makes it difficult to substitute a pretentious alphabetical soup of subscripted and superscripted notation for what is, essentially, our own ignorance.

It is not possible to present an entire introduction to Q-analysis here[6], because this essay would become too long; but I do wish to emphasize certain aspects of the methodology, and demonstrate the crucial links between the three perspectives of inquiry—a demonstration indicating that *all* are necessary, and that we choose one over another only at the price of impoverishing our descriptions, interpretations, and understanding.

The technical perspective

In the initial stages of a Q-analysis, there are clear and pressing reasons why the inquirer should bring the technical perspective to bear. Q-analysis is founded upon the definition of sets, and the definition of relations between sets. At first blush, both requirements appear naively obvious, and it is as easy to dismiss them as 'old hat' as it is difficult to grasp the deep implications and difficulties they pose for the sort of well-founded, replicable description that gains the assent of a reader by creating a plausible, but always contingent, structure. We define a set ... by *definition*, and we are obliged to rest upon such a tautology at these fundamental, rock-bottom stages. If pressed or challenged on a particular definition, we can only rest our case upon the principle of usefulness, which says that a definition is useful if it is useful—another tautology.

[6] Elementary introductions are contained in Atkin (1974a; 1981), and in Gould (1980). More technical presentations are available in Atkin (1974b; 1975; 1977a; 1977b; 1978).

Pressing further, a definition is useful if it provides the descriptive and factual substance of a meaningful interpretation. But such meaningfulness rests upon values and ideologies—perhaps examined or unexamined discourses—and we begin to see how, at this very first and fundamental stage, all three perspectives may be embedded in Q-methodology, just as Q-methodology is symbiotically embedded in the Habermasian troika.

We can define a set in two ways: by explication, that is by providing a list of all its elements; or by writing a rule that allows anyone to decide whether an element is, or is not, in the set. Sometimes the definition is quite easy; for example, the set of all countries in the Common Market. Sometimes it is extremely difficult; for example, the countries presently (1981) in the Arab League. If we are presented with an element, and cannot decide whether it is, or is not, a member of a set, we are forced back to the definitional task—which is precisely where we should be at the beginning. In brief, we must be able to decide; and if we cannot, we must examine carefully the question of whether we know what we are talking about. Many problems in the human sciences arise because sets have not been clearly and unambiguously defined, so people do not know what they are (literally) talking about.

But suppose an element, an elementary object of our observation, belongs to more than one set? Do we genuflect hurriedly towards Fuzzy Set Theory, and footnote Zadeh (1975)? The answer to the first question is that Q-analysis recognizes that elements may easily and obviously belong to more than one set, but such multiple membership raises questions of covers, which we will consider below. As for the second question, I have yet to see anything approaching an operational use of fuzzy set theory; I think it is totally unnecessary to invoke it for the inquisitive act (Haack, 1979); and I am increasingly impatient of footnotes and references implying that those who invoke them are working at the same level of nonoperational and disconnected abstraction as the cited authors.

As for defining relations between sets, I cannot help feeling that we have an example of the emancipatory perspective even at this initial stage of inquiry. For the simple fact is this: a relation is a highly unconstrained rule that assigns the elements of one set to another. In this apparently simple statement there is a provocative challenge; namely, to conceive of any meaningful definition of the word *explanation* which does not involve relating something to something else. To explain implies to relate: explanation cannot consist of something standing alone in the void. Connection of some things to other things is required, and this is why it is important to start with the most unconstrained form of a rule connecting things to things. The algebraic relation appears to be the most general one with which we can approach the descriptive task of inquiry.

I said the rule was unconstrained, because a function is a mapping is a relation (Gould, 1980). But the reverse is not true, and the implications for inquiry are deep. If we approach the descriptive task through the general rule of the relation, we will always find a function if one actually exists in the data sets. The reverse is not the case: if we look at the world through the tiny functional keyhole, then the world must conform to the highly constrained functional rule we are imposing. Unfortunately, we almost invariably approach the descriptive task in this way (think of the entire body of multivariate work in contemporary social science), and then usually end up by imposing the additional and severe restriction that the function must be linear. In brief, after looking through a small keyhole, we then insist we must only use a minute part of the aperture! In doing so, we crush complex multidimensional structure into 0-dimensional and highly-fragmented pieces. After such severe and damaging operations, it is a wonder that anything emerges through our linear sieves at all.

Why do we do it? Because we apparently do not know any better. And I think the reason we do not know any better is because in our search for more definitive and replicable knowledge we have followed quite blindly the paradigm of the physical sciences. In our quite understandable desire to be 'more scientific', an intellectual search totally comprehensible in the light of the previous inadequacies, we have looked to the physical sciences as our model of how we ought to proceed. But in the process of mapping our inquiries onto the structure of the physical sciences, we appear to have forgotten three things.

First, no physical science deals with the property of consciousness, a property that leads directly to the act of self-reflection (Gould, 1981b). Particles and elements do not argue back out of sheer perverseness, and the geometry that *is* the law of gravity appears to have no need for human comprehension or approval.

Second, in mapping our human geographic inquiries onto the paradigm of the physical sciences, we seem to have swallowed completely the highly constrained mathematical structures that are their fundamental languages (Gould, 1979d). Such languages traditionally assume the mathematically imaginative, but physically quite fictive, existence of the continuum, and they imply that algebraic operations upon the set of real numbers are well-defined and appropriate for the descriptive task. Inasmuch as numbers are invoked to measure and describe, these mathematical languages may be characterized as essentially quantitative.

It is important to acknowledge that such mathematical languages were initially generated *out of* the empirical substance of inquiry in the physical, chemical, and astronomical sciences, rather than being borrowed from somewhere else and imposed upon the observations. Moreover, they worked very well for several hundred years: there is no question that the function is a powerful descriptive device for an enormous range of physical phenomena. When the essential algebraic structure of these languages began to break down, new operations, leading to new and more appropriate structures, were devised. Men like Roger Penrose are honored in theoretical physics today because of their ability to modify and devise algebraic structures that can not only carry the weight of the empirical observations mapped onto them, but provide provocative leads about where to look next. In brief, it has become obvious that the particular physics one wishes to deal with depends crucially upon the structure of the algebraic language one chooses for the descriptive task (Atkin, 1965). I believe these more contemporary thoughts are precise reflections of what Heinrich Hertz (1956, page 4) meant when he said:

"By varying the choice of the propositions which we take as fundamental, we can give various representations to the principles of mechanics."

As for the concept of quantity and number, in the human sciences we assume the appropriateness of algebraic operations upon the set of real numbers R, even though we have to acknowledge that we *always* treat actual data in finite form[7], and can *never* actually observe an infinite number of those elements contained within the reals—for example $\sqrt{2}$ (Atkin, 1972). The point is neither pedantic nor trivial: well-defined algebraic operations on the reals do not transfer necessarily to the sets of the rationals and the integers[8]. Indeed, I think it is seldom allowable to use even the rationals, unless we acknowledge the perverse and unthinking stance of using a set of numbers when we can never observe and record most of them. In brief, virtually *all* our numerical observations are integers, or can be made so by simple scaling, and the algebraic operations we define must be appropriately constrained to the set of integers, Z. When we observe by counting (people, towns, places, and so on), this is obvious, but it is equally true when we think we have recorded the real number $13 \cdot 42173$ kilometers, instead of the integer number 1342173 centimeters.

Third, in our process of mapping our human inquiries onto the physical paradigm, we have innocently allowed ourselves to believe the ridiculous nonsense that statistics epitomizes the scientific method. So many in the human sciences appear to believe this today that even the statisticians are beginning to believe it themselves. Perhaps it shows the depths of our desperation in the late 'fifties to find something in human geography with

[7] We are extraordinarily perverse: even when we think data are recorded continuously (an aberration, that is to say an illusion, of the eye and brain, and the intermediary sensing instruments between the brain and events), we invariably break up the actual record into finite pieces for digital computation and our Fast Fourier Transforms. Even optical methods of computation are fixed for examination and permanent record on photographic film, which is then digitized again, or given *finite* numbers to measure gray scales that are only gray at some level of resolution where binary silver grains are spatially arrayed.

[8] As Bernard Marchand pointed out to me about ten years ago, when I was an even more innocent child than I am today. I now understand what his Gallic exasperation was trying to communicate in the context of our factor analytic discussions. The Anglo-Saxons are slow, but tenacious and affectionate—like the large dogs that roam the remote offshore islands of Europe.

an intellectual challenge to it. How otherwise can we explain the wild
Gadarene rush, when it is practically impossible to find a single major
scientific advance based upon the statistical approach? Kepler, Copernicus,
Newton, Darwin, Freud, Einstein, Crick ... there is not a statistic, not a
significance test, not a probability level among them[9]. As I have noted
elsewhere (Gould, 1979d, page 14):

"If you try to emulate the physicist of a century ago, and genuflect to
the statistician today, you are almost bound to start collecting numbers
and to end up falling into the conceptual trap of thinking that this is
what Science—and, therefore, social science—is all about ... if we
choose a mathematical language devised essentially for numbers, then
our apprehension of the world will be limited to numbers, and our
research will be constrained to those problems describable by numbers.
If human problems of significance wriggle away, and slip through the
wide meshes of our net, we should not be surprised—even if we are
chagrined when others point it out."

The initial and deep concern for rigorous set definition, and the
concommitant requirement of specifying relations between sets, force us
to acknowledge two further methodological tasks. The first is that the
elements of our sets are *words*, so immediately we find ourselves
enmeshed in that most human of all characteristics—language itself.
Q-analysis recognizes explicitly two things: that we choose words to
stand for things, and the words we choose to employ for our descriptions
may well exist at different levels of generality. If we choose words to
describe things, and we have a concern to create an area of publicly shared
and verified inquiry, then the definitions of words, and the meanings they
invoke or that we ascribe to them, must be agreed upon by all who use
them. Again, at first sight, this seems an obvious and even banal
requirement, one that human scientists should be able to take for granted.
But how many problems of human inquiry are obscure because different
meanings, different definitions, are given to the same words? How much
of published research in the human sciences is either an ambiguous and
multispecified mess, or consists of people talking past one another because
they have created their own private worlds of discourse by giving their
own meanings to words? If there is one thing the history of the physical
sciences can teach us, it is that shared bodies of understanding are created
when people start to care, passionately, for what words mean.
Communication of inquiry creates a body of shared, cumulative insight
when we mean the same thing when we use the same words. Otherwise
we are ships sailing in our own private worlds, failing even to dip our
flags in salute, or firing broadsides of misunderstanding at each other.

---

[9] Everyone is aware of the brilliant work of Boltzmann, and the equally pathbreaking
work of Einstein on Brownian movement. I do not equate such work on the
foundations of statistical mechanics with the rise of statistical methods.

Do you feel a terrible tension here? Even as you may be tempted to nod your head and say "Well yes, of course", do you feel an anarchistic glow of defiance, a secret desire to cheer on words, those sweet glories of our being, and rejoice in their ambiguity? Are you secretly on the side of the silver fish who slip through the net and will not be captured? I think such a tension dissolves when we acknowledge different domains in which we employ words. If we wish to work in the domain of inquiry, if we wish to share our inquiries and build a cumulative sense of understanding, then we must acknowledge the need to work with the hardest, most specific, least ambiguous definitions we can devise[10]. And if we do not agree upon definitions, then surely that is the place to begin the process of adversary challenge, to clarify *what* we are talking about. As a stage aside, let me note that the process I have labelled 'adversary challenge' in *Q*-analysis is *not* the bloodyminded, head-on, for-its-own-sake confrontation, where egos are bolstered, and loss of face is paramount. Rather, it is an open and welcomed process of criticism for the sake of clarification. Its motives are as pure and concerned, as deeply founded in *kundskapsvård* (Buttimer and Hägerstrand, 1979), as fragile human beings can make them.

Alternatively, if we wish to work in the domain of expression, either private expression, or expression inviting public response, then our words can only be unpinned; and those who use them must feel free to arrange and relate them for their own purposes, with all the ambiguous, multimeaning, punning, and metaphorical creativity at their command. The analogy I used above is not my own, so let me now acknowledge my debt. Here is Martin Seymor-Smith (1970, page 26):

"Language *is* a 'web', it is a network of relationships, that, for men, holds together a multitude of meanings that—like fish caught in a net —would otherwise swim apart."

And he is talking about one of the loveliest of poems by Robert Graves (1966)—the *Cool Web*. Here it is:

---

[10] At the same time, we must acknowledge the tautological nature of our definitions, in that words can only be defined by words, in the ultimately closed system we call a dictionary. This seems to be a fundamental tautology we have to live with, a tautology no different, in essence, than those upon which the physical sciences themselves are founded, where symbols stand for words to which meaning is ascribed by other symbols standing for words. F(orce) is *defined as* M(ass) $\times$ A(celleration); A(celleration) is symbolically stated as d^2s/dt^2, which occurs when a F(orce) is applied to a M(ass) ... the tautology comes out nicely in dimensional analysis, where we express the units as $F = MLT^{-2}$; $A = LT^{-2}$; and $M = M$. So $MLT^{-2} = MLT^{-2}$, or $MLT^{-2}/M = LT^{-2}$, which says acceleration is acceleration ... .

Children are dumb to say how hot the day is,
How hot the scent is of the summer rose,
How dreadful the black wastes of evening sky,
How dreadful the tall soldiers drumming by.

But we have speech, to blunt the angry day,
And speech, to dull the rose's cruel scent,
We spell away the overhanging night,
We spell away the soldiers and the fright.

There's a cool web of language winds us in,
Retreat from too much joy, or too much fear:
We grow sea-green at last and coldly die
In brininess and volubility.

But if we let our tongues lose self-possession,
Throwing off language and its watery clasp
Before our death, instead of when death comes,
Facing the wide glare of the children's day,
Facing the rose, the dark sky and the drums,
We shall go mad no doubt and die that way.

Yes, there is the tension: if we let the 'cool web of language wind us in
... we coldly die in brininess and volubility'. Words, especially those
pinned down by hard definition, and captured in a fine seined net from
which they cannot escape, such words do blunt, dull, and spell away the
heat, the scent, and the dread. *But*, yes, you see, here is the problem, if
we let our tongues lose self-possession, and throw off language, we shall
go mad, and die that way. We are dead if we do, and dead if we do not.
Unless, perhaps we acknowledge different domains of discourse and intent:
domains of inquiry, where the communication of shared understanding is
the intent; and domains of expression, where the intent is an invitation to
empathetic response. Peut-être ... qui sait?
   But in addition to its deep concern for language and the meaning of
words, Q-analysis also requires us to recognize that words exist at different
levels of generality, and to avoid logical difficulties we must sort out the
hierarchical structure of our words and our sets. Perhaps the easiest way
to appreciate this is with some examples. If we use, in the same breath,
with the same intention of specificity, the words Mathematics, Algebra,
and Geometry, it is clear that we are confusing levels of general meaning.
Mathematics is a higher level word, and contains, in a sense it is partially
defined by, Algebra and Geometry. We can think of {Mathematics} as a
set at, say, the $(N+2)$-level, whereas the sets {Algebra} and {Geometry},
themselves containing elements at a still lower hierarchical level, might be
at the $(N+1)$-level. If we use all three words as though they existed at
the same level, we are clearly confusing a set with the elements of the set.
Atkin (1974a), referencing Russell's "Theory of types" (Russell, 1956;

Marsh, 1956), warns us of the logical difficulties that lie ahead if we do this, although I am not sure that such a referencing is strictly needed[11].

Take another example: a dandelion is an object at some level, say N, which aggregates to the set {Flowers}, {Vegetables}, and {Weeds} at $(N+1)$ (Gould, 1980). Notice that the hierarchical structure is one of *cover* sets —the sets Flowers, Vegetables, and Weeds cover the element dandelion at the next level down. In brief, the hierarchical structures of language do not, in general, produce partitions—the bifurcating tree diagrams of the typical organizational chart. All partitions are covers, but not all covers are partitions. Gertrude Stein was quite right about her Roses, but felt constrained to operate at the N level only. When we acknowledge that a dandelion is a Flower is a Vegetable is a Weed, we note the existence of sets at the N and $(N+1)$ levels, and define a particular relation between them.

A final example comes from research on international television. Television programmes can be thought of both as content and as treatment, and sets of words have been defined at various levels to describe both of these aspects (Chapman and Johnson, 1979). In many cases, words describing the content of television programmes at the $(N+1)$-level aggregate to cover sets at the $(N+2)$, and do actually produce partitions. But some $(N+1)$-level words, such as Agriculture and Farming, aggregate both to the cover set {Environment} and to {Economic} at $(N+2)$, in the same way that Institutional Health Maintenance aggregates to the $(N+2)$-covers {Health} and {Welfare}. The basic problem in television research is that a programme can be about *anything*, so our first task is to draw up from the 'soup of everyday language', a primitive mixture of $N-?$ levels, a well-defined hierarchy of descriptive words, and state clearly the algebraic relations between them (Johnson, 1980a).

Two further methodological requirements are imposed by Q-analysis, and although they are definitional, and perhaps still largely within the technical tradition, they are clearly informed by the interpretive task to come. Again, we see the arbitrary nature of the division between the three perspectives, a partition forced upon Habermas only by the linear nature of narrative, rather than by any sharp conceptual separation between them. In Q-analysis a vitally important distinction is made between what

---

[11] I am indebted here to discussions with Roc Sandford, who refused to accept the necessity and relevance of the Barber's Paradox problem. Wittgenstein was also critical of Russell's Theory of Types, and his proposition 3·261 seems to be relevant here (Pears and McGinnis, 1961): Zwei Zeichen, ein durch Urzeichen definiertes, können nicht auf dieselbe Art und Weise bezeichen (Two signs cannot signify in the same manner if one is primitive and the other is defined by means of primitive signs). This appears to be close, perhaps identical, to the concept of an algebraic hierarchy of cover sets in language, in which terms at $(N+k+1)$ are defined by terms at $(N+k)$. In proposition 3·331 (and 3·332), Wittgenstein clearly rejects the Theory of Types, but I am extremely uneasy in this area.

is termed the *backcloth* and the *traffic*. These are unusual concepts when you first meet them, and at this point I could tease out an analogy, and try to draw parallels with independent and dependent variables. I am deliberately not going to do this, not because I do not want to communicate, or because I disparage the use of analogy, but because I believe that in this case it would ultimately be dysfunctional. It would tie our thinking to the old multivariate stuff, where we strain our observations through linear functional keyholes, and I deliberately want to break out of that trap. Let me try by giving you some examples instead.

A backcloth is defined by a relation (λ) between two sets, say F (farmers) and A (agricultural elements), and is usually denoted $\lambda \subseteq F \otimes A$. Think of a well-defined set of (F)armers as the rows of a matrix, strictly an incidence matrix Λ, and the set of (A)gricultural elements (land, tractors, co-ops, etc) as the columns—assuming, of course, that all a_i, $a_i \in A$, are at the same hierarchical level (Gaspar and Gould, 1981). Then each farmer can be considered as a simplex, a polyhedron whose vertices are those elements in the set A. The entire set of simplices forms a complex, strictly a simplicial complex, and I think you can see that if two or more farmers are characterized by similar agricultural elements they are connected through certain faces of their simplices. Simplifying a bit, the entire simplicial complex (as well as the conjugate structure—that is, agricultural simplices defined by farmer vertices), is the backcloth, and we can think of it literally in geometrical terms as forming a structure that can support certain things. The things that can exist on this structure are traffic, and in this case they are obviously crops, animals, fruit trees, and so on. Notice that the backcloth can exist without traffic, but that traffic *requires* a certain geometry to exist. No water, no crops—unless the backcloth is deliberately increased in dimensionality by augmenting the set A with irrigation. No tenured land, no investments in crops that require the security of a long-range planning horizon—for example, crops such as fruit trees, which may well be the most suitable for a changing agricultural region. Perhaps you can get a feel for the way the geometry of the backcloth allows or forbids certain forms of traffic to exist?

Or take something much closer to home: the research we do, the papers and books we write, the exercises and bibliographies we distribute in our teaching—all these things are traffic. But what is the backcloth, the supporting geometry? Perhaps a simplicial complex formed by the relation between an N-level set of faculty F [forming Department X at $(N+1)$?], and other sets such as {Secretaries, Graduate Assistants, Programmers, Librarians, ...}, or {Typewriters, Computer Terminals, Dictating Machines, Transcribers, Photographic Equipment, Duplicating Machines, Copiers, ...}. Does such a backcloth allow and forbid traffic? Ask those who must allocate hours and hours to these tasks, taking the time away from new traffic production. And hear the cries of anguish when the lack of maintenance funds means that computer terminals,

electric typewriters, and copying machines are torn out of the backcloth
and diminish its dimensionality! What stresses are felt when a gem of a
secretary leaves!

Or take television research again: the content of television programmes
is backcloth, a set of descriptive words, at some well-defined hierarchical
level, that form the vertices of programme-polyhedra. But traffic can be
anything that can exist on the underlying geometry so defined, not the
least being the way particular sets of subject matter are treated. For
example, a programme on the 0-simplex ⟨Dance⟩ may be treated as
Educational Time Slot, Child, Light Performance, Japan [12], and would
be a totally different programme from one treated with the traffic terms,
Academic Explanation, Academic Demonstration, Aesthetic Enrichment,
Serious Performance. The first is a light, gay, and joyous presentation,
perhaps inviting the participation of young children; the second is a
serious adult presentation to mature students of dance, with excerpts from
actual performances to demonstrate particular techniques. Actual
television programmes consist both of backcloth and traffic as mappings
are defined to 'glue' these two aspects together.

Or we may have a relation between a set of locations or factory sites X,
and a set of job skills Y, forming the simplicial complex $K_X(Y; \lambda)$ and its
conjugate $K_Y(X; \lambda^{-1})$. Trade unions tend to look at the complex, and
generate traffic such as wage claims on the skill vertices; while management
tries to consider how such changes in wage traffic translate on the conjugate
structure. As one side tends to put people first, while the other emphasizes
production, they are likely to be unaware of the way their negotiations
are switching from one geometry to another (Atkin, 1979a). It is intriguing
to think how the union of these two structures is made much more
explicit under conditions of worker-management. Indeed, and in a rough
sense, the hyphen in 'worker-management' stands for the amalgamation of
viewpoint, the merging of perspective when such social and economic
experiments are tried.

As a final example, consider an experiment I try every year in a small
graduate course. I first ask people to list, in a totally open and unconstrained
way, their intellectual interests, and then ask them to arrange these in a
hierarchy of cover sets. This is an intellectually demanding task: there is
usually blood all over the walls after one week, and some people have
actually dropped the course in a high emotional dudgeon, saying the task
"can't be done", it is totally unreasonable, ridiculous, etc. It is interesting
how problems concerning the definition and meaning of words can raise
emotions so quickly, and such stress clearly indicates we are asking people
to deal with something at the core of their own humanity. Whether we
find a satisfactory hierarchy of cover sets or not, we usually examine the

[12] Three such programmes were broadcast on 12 September 1977, by the educational
channel of NHK (Nippon Hoso Kyokai), the Japanese Broadcasting Corporation (Gould
and Sugiura, 1980).

relation $\lambda \subseteq P \otimes I$ for pedagogic purposes, and I give each p_i who is a member of the set P a *nom de plume*, so we can talk about one another without embarrassment, or without a feeling that we are invading each other's privacy[13].

The complex $K_P(I; \lambda)$ generates a backcloth, a geometry in which the people are defined and connected by their intellectual interests. This raises explicitly the traffic that could exist on such a geometry—traffic such as ideas requiring particular vertices, and needing shared faces of a certain dimensionality before they could be transmitted (Johnson, 1980b). Such a description also allows for thoughtful serendipity effects: for example, the intriguing fact that in one seminar the connection between the conjugate simplices Structure and Poetry turned out to be the face and simplex Music. If the former represents an intellectual concern for rule-determined connection and form, whereas the latter indicates a concern for rhythm and pattern, then the face of these might well define Music, in some constrained and limited sense. Moreover, if people evaluate their knowledge of certain terms and concepts at the beginning of the year, and then are asked to repeat the experiment at the end, it is possible to see how they have increased their 'intellectual dimensionality'. If the seminar has been a success, they should now be much more highly connected, and so capable of working with, and transmitting among themselves, much higher dimensional ideas and concepts.

And notice two things: all the discussion of the past few pages, all the examples from television to agriculture to management to teaching, have been concerned with relations between *words*, and the interpretation of the structures so defined. There has not been a number in sight. Do you see now Helen Couclelis's (1982) concern for a *qualitative* mathematics? Second, the focus upon definition, meaning, and hierarchical level of generality makes us much more sensitive to words themselves. What do such words as Democracy, Liberty, Freedom, Welfare ... mean? They are so high in the hierarchy at $(N+7)$, and so many words with much more specific meanings at $(N-1)$ aggregate to them (according to relations that are undefined, but clearly multiple and private), that such words can mean all things to all people. This is why they are beloved by politicians at election time, and why their meanings are so easily twisted and perverted: you can read anything you like into them, and aggregate any $(N-1)$-terms with your own, quite personally defined, relation. When the elections are over, and the international agreements are signed, the aggregate ambiguity allows a thousand reasons for failing to keep the promises they seemed to contain. As a general rule, politicians and negotiators do not like being

[13] When one set is defined by people, you have to be extremely careful—even for pedagogic purposes. Q-analysis tends to approach the psychoanalytic at these levels, and a deep respect for individual privacy within a group must be observed. I have found the same care is necessary in a study of 'tactical space', an examination of the topology of connections linking tactics people employ to 'get their way' with others.

pulled down the hierarchy where words take on more precise and specific meanings.

Before leaving these questions of language and qualitative mathematics—questions rooted in the technical perspective, but clearly informed by the hermeneutic tasks to come—I want to speculate about some possible parallels which may exist between the methodological perspective of Q-analysis and the thinking of Wittgenstein. In his *early* concern for delimiting the bounds of the sayable, the "bubble of discourse", he put forward, essentially and literally, a pictorial view of language in which propositions were composed of elementary sentences correlating names with objects. Objects can be related in a very large, but still finite, number of ways, and when the names we define are related in a one-to-one correspondence with the objects, we may speak of a true proposition taken from the set of all possible arrangements. In terms of Q-analysis, I cannot help feeling that a relation, a particular λ we define, is identically equal to a proposition of the early Wittgenstein, in other words $\lambda \equiv$ Proposition. We define objects in terms of the vertices contained in another set, and the relations between our objects define a structure. Making the concept of structure operational in this way (a way that is transcendentally pictorial even at high, multidimensional levels) seems to reflect Wittgenstein's idea that the world must have a certain structure in order for it to be represented in language at all. I said 'represented' because I think this comes closer to his idea that the attachment of language to the world, of the word to reality, can be shown but not stated, at least not in verbal languages, although possibly in the graphical and the algebraic. I refer specifically to Wittgenstein's own eureka experience when he saw a map in a newspaper.

The early Wittgenstein's deep belief that our ability to talk about the world rests upon our ability to describe the particular facts that compose it helps us understand an aspect of Q-analysis about which many have feelings of distinct disquiet. A relation λ may be stated as an incidence matrix Λ defining the binary relation between the elements of sets. Yet the initial phase of a Q-analysis is by no means confined to the collection of simple 0–1 statements; on the contrary, we may have integer values representing intensities of relationship between elements, and then choose slicing parameters that transform, that literally map, our more conventional 'data matrix' onto the set $\{0, 1\}$. For example, farmers may rank those to whom they turn for agricultural advice; foraminifera may appear in undersea grab samples at different degrees of intensity; people may evaluate the degree of their interest or competence in particular subjects, and so on. We may choose a single slicing parameter θ, a slicing vector θ_i or θ_j, or even a slicing matrix $\theta_{ij} > \Psi$. We can slice globally on all elements, or locally on a single element. We can define our relations by 'slicing out' high values, low values, or by choosing a 'middle cut' if we so choose—for example, an agricultural structure defined by farmers between forty and fifty years old. In each case, the θ chosen defines our λ, and the simplicial complex and its conjugate defined by this binary relation offer a particular *proposition* about the world.

Is not such a proposition arbitrary? Does it have a truth value in that there is a one-to-one correspondence between it and the world? To answer such questions properly, I think we have to turn to the later Wittgenstein of the *Philosophical Investigations*. The proposition is not arbitrary; on the contrary, it is well and clearly defined, a structure chosen out of what must be considered, for all practical purposes, an infinitude of possible structures[14]. The choice is, of course, predicated on the assumption that we have observed correctly, and that the data are as secure and well-founded as we can make them. But equally important is the question as to whether the proposition we have put forward has truth value, and this can only rest upon the interpretation given to the particular structure we have defined. The basic question here is whether the interpretation has meaning, and meaning is ultimately rooted in the usefulness of the interpretation for our investigative task. But meaning, and thus agreement about the usefulness of one of our 'arbitrary' structural propositions, rests upon two prior considerations—the degree to which we share the rules of language (the degree to which we can use language as a common tool), and the degree to which we share similar value structures. In the inquisitive use of language we are playing a particular language game—and recall that Wittgenstein was deadly serious about the game

[14] Consider a small example, perhaps a relation defining a seminar with eight students and fifty intellectual interests. Then the number of potential structural propositions we can describe and offer for interpretation will be 2^{400} or more than all the electrons in the universe. Faced with such unthinkable numbers of combinatorial possibilities, it is surely tempting to shift to the entropy mode, and ask what is the most likely state of the system. But we are dealing with people here, and what meaning can such a statement as "most likely state" have in this context? I remember, in 1969, coming across the work of Per Martin-Löf (1969–70), and becoming particularly enamoured with a similar example of children versus tests. A 0–1 matrix describes which child passed which test, and if we assume, in the absence of all other information, that this is the "most likely state", then iterative procedures allow us to compute essentially a Boltzmann function in which a parameter characterizes each child and test. These might be interpreted as indices of child 'test capability' and test 'difficulty'. But quite apart from the fact that such indices would probably correspond closely to simple row and column totals, what a poverty-stricken description we are left with! All the multidimensional richness, all the detailed insight, which is presented to us in $K_C(T; \lambda)$ and $K_T(C; \lambda^1)$ has been crushed down to a single number, and even this is meaningless because test results are clearly 0–1 *traffic* on a much more humanly relevant backcloth, where vertices defining the children would require extremely careful and thoughtful definition. This is why I feel such multivariate and statistical *mechanical* approaches, which crush relational information away with the function 'press', can never give us the same deep interpretive insight into human affairs as Q-analysis, and can never provide the same opportunities for those serendipity effects that come from staying close to the data through often long and patient exercises of interpretive analysis.

analogy, emphasizing the purposeful, rule-governed nature of games, rather
than their recreational, perhaps frivolous, nature. Interpretation via shared-
language games has taken us to the hermeneutic perspective, even as the
knowledge of shared values has taken us to the emancipatory.

As a final speculation, I cannot help mapping much of the early
Wittgenstein of the *Tractatus* onto the technical perspective, and relating
much of the thinking in the later *Philosophical Investigations* to the
hermeneutic, with both mappings consciously informed by an emancipatory
concern. Despite the criticism contained in the latter for some of the
ideas in the former, such a view emphasizes the essential continuity of his
work and thinking. And since I am emphasizing structural correspondences
between acts of philosophical investigation by different people—acts which
may appear superficially different, but which may actually correspond at a
deeper level—I cannot help wondering if category theory may not become
a suitable algebraic language for describing such assignments, that is
functors, from one category of investigation to another (Hilton, 1969;
Goldblatt, 1979; Arbib and Manes, 1975). Perhaps this is close to what
Stafford Beer had in mind when he spoke of mapping a value system from
one culture onto, or into, another (Beer, 1976).

As for the emancipatory tradition and its correspondence, I am still a
little worried about what will happen if I kick my own q-ladder away.

The hermeneutic interpretation of structure
Suppose we have well and truly defined our sets, and we have examined
and decided upon cover sets and hierarchical relationships. Suppose we
have carefully distinguished between backcloth and traffic, and chosen a
slicing parameter to define a relation between the sets. At this point,
we have created a particular structure, represented by our complex and
conjugate.

But what does it mean?

The meaning we give to such a structure is founded upon the act of
interpretation. The geometry we have created is a space, a backcloth against
which existence and action can, or *cannot*, take place. A fundamental
consideration is the structure, the multidimensional connectivity, of the
topological space created by the relation[15]. The degree of connection
will vary with the dimensional level we consider: part of Q-analysis
consists of putting on spectacles whose lenses allow us to 'see' structures
at successively lower dimensions. To give a concrete example, a farmer-
simplex may be defined in terms of other farmer-vertices who approach

[15] *Strictly*, a simplicial complex is not a topological space, although each simplex
represents a discrete topology. The underlying point set of a simplicial complex K
can be considered to have a subset topology, denoted $|K|$, as the union of all simplices
in K. Thus, we can assign every K its underlying topological space $|K|$, and in fact
$K \rightarrow |K|$ is a functor from the category of simplicial complexes to the category $\mathcal{T}$ of
topological spaces (Hilton, 1969, pages 83-97).

him for agricultural advice. At high q-levels we shall see only very high-dimensional farmer-polyhedra, and these may not be connected by high-dimensional faces. At these levels, in other words looking through these high-dimensional lenses, the backcloth may be very fragmented, and so offer a high degree of obstruction to the transmission of traffic—in this case, new ideas, innovations in farming, and so on. An obstruction vector $\hat{Q}$ gives us, for successively lower dimensions, a measure of the 'gaps' in the backcloth offering obstruction to the transmission of traffic.

The obstruction vector provides what we might call a *global* view of connectivity in our geometrical space, but we may also require a much more detailed, local view. Within a component (one or more connected simplices in a portion of a fragmented backcloth) the connectivity is represented by a q-chain. But this could vary from one extreme, where all the simplices (think about farmers) are strung out to a chain of maximum length, to the other extreme where all are connected through a single face to form a star. Such local structure will clearly influence the transmission of traffic on each part of the fragmented backcloth. This view, founded upon the definition of a relation, throws a devastating light on the estimation of a Mean Information Field in conventional diffusion theory. Multidimensional complexity, and extreme variation in local structure, are simply crushed out and obliterated by the functional approach.

Measures of eccentricity can also be defined, and they conform well to the intuitive and common use of the term. A simplex sharing all its vertices with others in the complex is 'right in the middle of things', conforming to shared attributes, and displaying zero eccentricity. In contrast, a simplex defined by a number of vertices unique to itself 'sticks out like a sore thumb', and may be considered highly eccentric.

In addition, the spaces we define may exhibit certain properties which can only be described as holes—Q-holes, since they exist at various dimensional or q-levels. A curious property of such holes, and a considerable interpretative challenge, is that they act as objects in our usual spaces, rather than 'absences' in our conventional thinking. In topological terms, a tree or a large boulder is a hole in conventional E^3, and since we are unable to pass through such holes, we have to work our way around them. These 'topological objects' have been found in committee structures in universities (Atkin, 1977a), and were intuitively felt as obstructions around which memoranda and report traffic moved in a 'buck passing' fashion. They have also turned up in an analysis of a fugue, presumably created by the rules of transition determined both by the key as well as by the musical conventions of the composer's day. It is provocative to think of Schönberg (and his students Webern and Berg) defining new relations on a set of notes, so changing and enlarging the geometry of the backcloth to allow traffic which was previously forbidden[16].

As a matter of fact, we have all experienced the obstructive nature of q-holes in some of the geometries which form the backcloths for our own lives—particularly in bureaucracies, where people and forms are shuffled from one in-tray to another, and where decisions never seem to be taken at the $(N+1)$ level where we find ourselves. As a result, we often experience the frustration of trying to invoke the sympathy of the q-hole fillers who exist at the $(N+2)$-level—people who are only reachable via well-defined relations called 'appropriate channels' (and don't you forget it!).

Notice that in these interpretative tasks there are no guidelines—no handles to crank, no residual glow to name after tons of statistical pitchblende have been crushed through our linear sieves. What is required is a patient hermeneutic process of interpretation that depends upon imaginative insight and a deep knowledge of the substantive area and particular problem, a process aided by staying as close as possible to the data. In a sense, we are presented with a well-defined structural *text*, and are then asked to bring to bear all the interpretative insight we can to elucidate meaning. Of course, a particular structure may mean nothing, either because we do not have the perspicacity and interpretive skill, or because the sets and relations were badly or inappropriately defined in the first place. I have had students, sometimes utterly frustrated and quite angry students, complain bitterly that Q-analysis had let them down, when in fact their own abilities had been shown up. Q-analysis is often a circular process, a process of definition, analysis, interpretation (or noninterpretation) leading to redefinition, reinterpretation, until things make sense, until 'things feel right'.

Such a process is science? Yes, of course: what characterizes scientific inquiry unless it is careful definition, scrupulous observation, and a deep respect for the facts of data, patient interpretation, reexamination when things 'don't make any sense', redefinition ... until a convincing story can be pieced together?

And beyond these perspectives, beyond the insistence that is *structure* that holds the key, Q-analysis provides a language that allows us to describe things as they are. In brief, it calls for a mathematical structure that is genuinely robust, and much more capable of carrying the observations we make, rather than mapping them onto constrained algebraic forms that force the human world to conform to restricted functions. It also opens up views on process and change that we have not really considered before.

---

[16] Forte (1973) has developed a set theoretic description of 20th Century music which might well be extended by more explicit topological concepts; and the writing of Rosen (1976) on Schönberg is saturated with structural description which might be made operational.

We have, in essence, two views of change: change in traffic that exists on the backcloth, and change in the backcloth itself. The former is perhaps a Newtonian view; change is a result of forces—forces of attraction and forces of repulsion—that alter the things that exist on the underlying structure. In that we measure forces by observing change (in extension, in velocity, in direction, etc), so we ascribe change to forces—they are the same thing. Such forces are graded, which implies they are associated with the particular dimension of the simplex supporting the traffic. Since a mapping of the sort:

$$\Pi = \pi^0 \oplus \pi^1 \oplus \dots \pi^t \oplus \dots \pi^n \ ,$$

defines the graded pattern or traffic on various pieces of a simplicial complex, such forces are usually termed t-forces. For example, a recession is felt by employees (traffic) on a backcloth that may be defined by a relation between a set of places in East Anglia and a set of employment categories. Forces of repulsion are far stronger for females than for males, as this segment of the traffic is crushed out by the t-forces operating on various graded portions of the backcloth (Atkin, 1977c).

But we are also offered an alternative view of change, a view we might term the Einsteinian view, where events, and changes in events, are a result of what the geometry allows. In the physical sciences it is assumed that the backcloth is stable and unchanging; but this view will not hold in the human realm, where the backcloth itself may change, and so alter the traffic which exists upon it. The differential, or the finite difference, denoting a rate of change in conventional notation, is no longer conceptually adequate to handle the sort of change experienced in the area of human inquiry, since the differential refers only to a change in traffic against a *stable* backcloth. Change from t_1 to t_2 can be represented properly (that is, without mapping the observations onto an inadequate mathematical structure) only by a strain pair recording both backcloth and traffic changes by the difference between pattern polynomials—the algebraic expressions of the geometrical spaces we have created (Johnson, 1975; 1977).

Even deeper possibilities are opened by this methodological perspective on human inquiry, for the notion of time itself is enlarged. In the discussion above, I have assumed implicitly that our notion of time is the Newtonian view of the 'ever-moving stream', symbolized by t and incorporated as such into all differential expressions involving change. But such a view of time implies a particularly simple structure of events consisting of zero-simplices connected to past and future event-simplices in a complex in which time moments and time intervals are the graded patterns, $\tau = \tau^0 \oplus \tau^1$. Such a view no longer holds in a relativistic view, where the space–time continuum must be represented as a graded pattern, $\tau = \tau^3 \oplus \tau^4$, so that moment events are now graded up to three dimensions, while intervals take gradations up to four.

However, when we move onto backcloths appropriate for the description of human and social phenomena, time becomes a traffic which totally orders all the simplices in the complex. We experience time as a $0-1$ function on the simplices, but we often fail to realize that the movement from one simplex to another must be controlled by the connectivity of the backcloth. One event of a certain dimensionality, say p, cannot come after another unless there is sufficient connective tissue, namely a $(p+1)$-interval—in exactly the same way that a 0-dimensional 'now event' is connected to the next by a 1-dimensional interval in linear Newtonian time. This means our 'sense of time' will be conditioned by the particular structure in which we find ourselves, or on which we live as traffic. If we *assume* a bijective mapping from structural time to Newtonian time, then we may expect events of varying dimensionality to appear at varying Newtonian time intervals—approximately three days for a 1-event, up to 351 days (approximately one year) for a 25-event. But Newtonian time is cyclical time as year succeeds to year, and so produces a q-hole. To quote Atkin (1978):

"... this means that scientific (cyclic) time is noise ... . In comparison, the time-traffic we experience in our personal structure ... is silent traffic. It corresponds to the experience of achievement, of working through the structure. The noise of scientific time is a nuisance, by comparison."

It is difficult to convey some of these perspectives without considerable prior explication, but I think we can get an intuitive sense of what these notions imply both at the individual and at the societal levels. At the level of the person, I think Jung's views on individuation can be seen from a rather different perspective. Individuation is, in essence, a process of personal growth, which can be considered as an augmentation and enlargement of an individual's dimensionality. Such a view implies certain paths chosen; a steering of the now-individual to the future-individual by the acquisition of new vertices. For example, if I want to enlarge my life by learning to love Swedish poetry, I must acquire the vertex ⟨Swedish⟩ at some high slicing level of competence so that the I-polyhedron is enlarged. If I want to understand category theory, the now-I must acquire the vertices ⟨algebra⟩ and ⟨topology⟩, for only then will the connective tissue to the future-I be capable of carrying the $\{0, 1\}$ time traffic. If we want to grow, to increase our dimensionality, we must make choices now to become. But do you see how we have shifted at this point to the emancipatory perspective?

As for the societal level, what does such a multidimensional perspective on time imply for planning, for the specification and reaching of goals? Some goals are small p-events, easily reachable by the $0-1$ function, because sufficient $(p+1)$ connective tissue already exists. But other goals, much higher dimensional p-events, require the construction of connective tissue in the topological structures of our multidimensional world. The future-society requires the now-society to act, to choose, to construct the vertices that allow the 0–1 function of time to move on the backcloth. In this way, the structural view of time makes explicit the necessity to steer, to act, to *plan* to reach those *p*-events valued as more humane and more desirable. Once again, we have landed up in the emancipatory perspective, for the next question must be *what* goals and *what* values will inform the choices that create the connective tissue to the future we want (Gould, 1977b). And, equally important, are there the resources (intellectual, technical, and material) to make the choices? Do such resource-vertices exist to build the structures of *p*-events—the event-structures of particular geometries which will allow, in turn, the traffic we desire?

And it is here that we connect with the whole decision process, a topic so inadequately and naively treated by the statistical approach over the past forty years. For decisionmaking is nothing more than the attempt to select from the now-horizon certain *q*-events: events not on a partitional tree with 0-event vertices as a backcloth for our 0–1 choice function, but on a multidimensional 'cover structure' which recognizes the *p*-event nature of possible futures, and allows us to see what is required *now* to forge the connective intervals to *later* (Atkin, 1979b).

The emancipatory structural perspective

Parts of this essay have refused to be confined to the three perspectives of inquiry, making it obvious that all three are inextricably linked together as each informs the other. It is not a matter of choosing one, but realizing that the structural approach to the descriptive task requires all of them. My final task is to demonstrate that *Q*-methodology is rooted in the emancipatory tradition, quite as much as it is founded upon the technical, and informed by the hermeneutic.

If we adopt the perspective of *Q*-analysis, our thinking is immediately conditioned by the geometric and structural viewpoint, and the realization that we live in multidimensional spaces that both allow and forbid. If we crush these multidimensional geometries down, by destroying them with functional estimating procedures, and by projecting these multidimensional worlds onto the flat maps that are our traditional cartographic tools[17], we can only gain a diminished, perhaps poverty-stricken and highly distorted view of the world and ourselves in the world.

[17] Do not misunderstand me: no one is more respectful or intellectually more delighted by map representation, and I have advocated repeatedly the extension of the cartographic perspective, and the enlarged definition of the map, to include any graphic representation of relations between elements of a set. Such graphic pictures of unusual 'spaces' can be enormously thought-provoking, and help people to say literally "I see"! But we must use them with care, realizing that the complexity of the multi-dimensional spaces we describe cannot always be expressed in our graphic language, but may have to be conveyed by the algebraic.

But such descriptions also raise the question of changing the geometry, for the structural view makes it clear that our backcloths are *not* unchanging, so they are by no means deterministic in a perpetuating sense. This is why there are no laws of human behavior, for we have the ability to get outside and look in, and so alter the behavior that *is* traffic, as well as the backcloth structures that allow and forbid behavior. Revolutions—American, French, and Russian—change the topology, alter the backcloth, and so allow new traffic. The only question is whether the new backcloth is enlarged or diminished—and in what ways (Marchand, 1974).

I think it is because Q-analysis provides such perspectives that the philosopher Melville (1976) has termed it the first example of an emancipatory mathematics in the social sciences. Not only does it allow a description of structure that is rooted in actual observation—a description that refuses to crush the data with inappropriate and constrained mathematical forms—but it also raises the possibility of changing the structures themselves. Thus, there are no laws; although Atkin (1981) has raised the question of metalaws of human behavior, laws (shall we say 'regularities'?) which indicate how the laws (shall we still say 'regularities'?) themselves change. And here we see the never-ending recursive process that consciousness provides (Hofstadter, 1979): a process of contemplating ourselves contemplating ourselves contemplating ... . Perhaps the universe produced consciousness in order to contemplate itself (Wheeler and Patton, 1977)?

The ability of Q-analysis to help us reach an emancipatory perspective (and perhaps a metaperspective to view Q-analysis itself?) is not confined to releasing us from the view that human inquiry is a search for laws. As a language of structure, written in the notation of algebraic topology, it is a very general, high-level mathematics providing us with a perspective on more constrained approaches: the traditional differential is only a special notation for particularly simple and constrained forms of change against a stable, unvarying backcloth; the function is a highly constrained form of a relation; the partition is a special case of the cover set, and so on. The consequences are devastating for almost all multivariate work in the social sciences today, and liberating in that we are no longer constrained to think in these limited ways. Just think of all the factor analytic approaches, and the way they start by crushing the detail away with a linear correlation coefficient. As Marchand (1978; 1979) has noted, after hundreds of urban factorial ecologies, we have discovered the structure of the US Census. Recent critiques of such analyses, in which data sets are reanalyzed by Q-analysis, indicate what a poverty-stricken approach such multivariate procedures are (Chapman, 1981; Gatrell, 1981; Gorenflo, 1982).

Recall, too, that such multivariate approaches form the basis for most work in numerical taxonomy, and for most regionalization and classification today in geography. Most numerical taxonomies start from a consideration of two sets, a set of objects to be pigeonholed, and a set of variables (usually a set of redundant and ill-specified elements existing at all sorts of levels in an unspecified hierarchy). The whole lot is squeezed through a linear filter, often by performing totally undefined algebraic operations (adding ranks, taking means, etc), so that a configuration of points is provided in a metric space, even when the data were originally nonmetric to begin with. Some algorithm, *any* algorithm, or a multiplicity of algorithms, is then let loose in the space to group the points (objects) together. Different algorithms produce different groups and taxonomic trees, and if the results still do not conform to prior expectations, more constraints are added until they do. The whole data set is forced through a deterministic partitional machine which, *by definition*, must produce partitions on the set. There is nothing here that even faintly resembles science. People in the humanities know better. Here, after a lifetime spent in the world of words, is James Murray of the *Oxford English Dictionary* (Murray, 1979):

"How often does a man hairsplit, and sever, and part asunder what Heaven has made a whole! ... man is fond to classify, to separate, to discriminate, to set apart in little cells of memory the mass of facts he gathers from the field of nature. But nature has no such isolating methods—her facts and laws are a continuous, all-connecting network."

In contrast, a Q-analysis lets the 'data speak for themselves' (Gould, 1981b), and is concerned to explore and interrogate structure, *not* tear apart a geometry of connective tissue to stuff the pieces in little partitional boxes to satisfy what one distinguished virologist has termed "the potty training of the observer". It is precisely the emancipation from such potty training, such engrained habits of thought, that a structural approach provides; and in the whole process it raises the question of whether we should discard the 200-year-old tradition of Linnaean thinking that lies behind it. Perhaps Bonnet and Buffon were right after all (Foucault, 1973)? And notice that if partitions do actually exist at some well-specified dimensional level, then a q-analysis will find them—all partitions are covers—but it will not force a partition upon a well-connected structure, tearing it apart to satisfy the deterministic potty training of a Linnaean thinker. For example, in a reexamination of a data set of foraminifera, I found that the conventional 'intuitive' approach of an older generation, and the newer, numerical taxonomic approach of a younger generation, were both capable of doing precisely the same damage (Gould, 1981a; Craig and Labovitz, 1981). None of the postulated 'assemblages' existed in the data; all were figments of the constrained thinking which was applied to the investigative task. I have had competent geologists and oceanographers confess, after cutting 'assemblages' from seven to five to two during a doctoral examination, that none were sure about such categories even at the end of it all.

In television research, we have found that data on programmes were labelled 'ambiguous' and 'intractable', a lovely anthropocentric view of an inanimate data set that was doing its best to tell the partitional thinkers something about the well-connected *structure* of television programming (Gould, 1978). Because of Q-analysis, we have now adopted the view that it is much more appropriate to help people define, explore, and interrogate television structures, rather than help them put these complex and highly connected artifacts of contemporary human culture into 18th Century, nonoverlapping boxes. And since we now have computers to augment our intelligence, we might as well start using them as prosthetic devices instead of just big adding machines, only doing faster what we did in the same ruts before.

And here is the final aspect of the emancipatory view: so much of our thinking tends to run in channels—channels continually eroded by rivers of thought that cut deeper and deeper in the same rut. You can see such thinking everywhere: in years of stolid inquiry in the same direction that seems to lead nowhere, and in 'new' journals that reflect the same thinking of the editors when they were graduate students. How refreshing to hear Herbert Simon (1980) testify before a Congressional Subcommittee that "a shift in structure is not anything that is easily accommodated in the kind of economic theories that we have now". Perhaps one day we shall have a structural description of the economic backcloth, and realize that most of the indices so assiduously collected are but traffic on a complex structure. At the same time, how sad to see the deeply incised rut of thinking represented by three decades of research on chess, an ever-extending thought-pattern of tree-searching, augmented by 'look ups' of positions (Simon, 1979), when we know Grand Masters do *not* think in these N-level tactical terms, but the $(N+1)$-level terms of positional chess that are *essentially* structural (Atkin and Witten, 1975).

Finally, Q-analysis can help us to obtain an emancipatory view of things deeply human. For example, much concern is presently being shown for problems of gender, and research programmes are being proposed that will extend our knowledge, insight, and, above all, our understanding beyond the first, and still necessary consciousness-raising stages that came mainly from the Women's Liberation Movement. The literature on gender is redolent with structural terms, many of which might be made more precise and operational through the language of Q-analysis.

Many women in academia, for example, may well wonder what geometrical backcloth they are living on—the officially sanctioned one, or the unofficial, and often hidden, chauvinist structure (Laws, 1975). Whether gender is defined physiologically, in terms of self-identity, or by the ascription of gender to an individual by the society, it is clear that it is *not* a partitional, either–or concept (Kessler and McKenna, 1978), and we do great damage to ourselves and others by forcing people into the conventional dichotomous boxes. People are polyhedra of varying dimensionality, and inasmuch as we constrain children to certain vertices, we limit their humanity. Moreover, our androgynous selves change from society to society, and from one time of life to another (Bem, 1974). I believe, with David Mulhall (1977) at Middlesex Psychiatric Hospital, that the appropriate structural description may well provide a more human, helpful, and emancipatory view.

Let us, therefore, take an emancipatory stance, and rather than choosing one perspective to the exclusion of others, or juxtaposing one examined ideology in direct confrontation with another, "descend", with Martin Heidegger, "once again into the poverty of [geography's] materials" (Magee, 1978). And there, exploring, let us see what Common Ground we can find.

References

Ackoff R, 1978 *Art of Problem Solving* (John Wiley, New York)
Arbib M, Manes E, 1975 *Arrows, Structures, and Functors: The Categorical Imperative* (Academic Press, New York)
Ashby R, 1956 *Introduction to Cybernetics* (Chapman and Hall, London)
Atkin R, 1965 "Abstract physics" *Il Nuevo Cimento* **38** 496-517
Atkin R, 1972 "From cohomology in physics to Q-connectivity in social science" *International Journal of Man-Machine Studies* **4** 139-167
Atkin R, 1974a *Mathematical Structure in Human Affairs* (Heinemann Educational Books, London)
Atkin R, 1974b "An approach to structure in architectural and urban design. 1. Introduction and mathematical theory; 2. Algebraic representation and local structure" *Environment and Planning B* **1** 51-67; 173-191
Atkin R, 1975 "An approach to structure in architectural and urban design. 3. Illustrative examples" *Environment and Planning B* **2** 21-57
Atkin R, 1977a *Combinatorial Connectivities in Social Systems: An Application of Simplicial Complex Structures to the Study of Large Organizations* (Birkhäuser, Basel)
Atkin R, 1977b "Methodology of Q-analysis" Research Report IX, Department of Mathematics, University of Essex, Colchester
Atkin R, 1977c "Q-analysis: theory and practice" Research Report X, Department of Mathematics, University of Essex, Colchester
Atkin R, 1978 "Time as a pattern on a multidimensional backcloth" *Journal of Social and Biological Structures* **1** 281-295
Atkin R, 1979a "How to study corporations by using concepts of connectivity" Department of Mathematics, University of Essex, Colchester (mimeo) 17 pages
Atkin R, 1979b "A kinematics for decision making" NATO Conference on Applied General Systems Research, State University of New York, Binghamton; available from Department of Mathematics, University of Essex, Colchester
Atkin R, 1981 *Multidimensional Man* (Penguin Books, Harmondsworth, Middx)
Atkin R, Casti J, 1977 *Polyhedral Dynamics and the Geometry of Systems* (International Institute for Applied Systems Analysis, Laxenburg)
Atkin R, Witten I, 1975 "A multidimensional approach to positional chess" *International Journal of Man-Machine Studies* **7** 727-750
Barthes R, 1957 *Mythologies* (Editions du Seuil, Paris)
Barthes R, 1963 *Sur Racine* (Editions du Seuil, Paris)
Barthes R, 1977 *Fragments d'un Discours Amoureux* (Editions du Seuil, Paris)
Beer S, 1976 *Platform for Change* (John Wiley, New York)
Bem A, 1974 "The measurement of psychological androgyny" *Journal of Consulting and Clinical Psychology* **42** 155-162
Buttimer A, Hägerstrand R, 1979 *Invitation to Dialogue: A Progress Report* DIA Paper number 1, Department of Geography, University of Lund, Lund

Chapman G, 1981 "Lists of capes and bays: and of towns and companies and schools
    and ..." in *British Quantitative Geography, Prospect and Retrospect* Eds R Bennett,
    N Wrigley (Routledge and Kegan Paul, Henley-on-Thames, Oxon) pp 1-24
Chapman G, Johnson J, 1979 *Television Programme Coding Manual* International
    Television Flows Project Report number 6, Department of Geography, Cambridge
    University, Cambridge
Chomsky N, 1968 *Language and Mind* (Harcourt, Brace and World, New York)
Couclelis H, 1981 "Philosophy in the construction of geographic reality" this volume,
    pp 105-138
Craig R, Labovitz M, 1981 *Future Trends in Geomathematics* (Pion, London)
Deskins D, 1977 *Geographic Humanism, Analysis and Social Action* Publication 17,
    Department of Geography, University of Michigan, Ann Arbor, Mich.
Eriksson E, 1958 *Young Man Luther* (W Norton, New York)
Ford L, Fulkerson D, 1962 *Flows in Networks* (Princeton University Press, Princeton,
    NJ)
Forrester J, 1971 *World Dynamics* (Wright-Allen Press, Cambridge, Mass)
Forte A, 1973 *The Structure of Atonal Music* (Yale University Press, New Haven, Conn.)
Foucault M, 1973 *Les Mots et Les Choses* translated as *The Order of Things: An
    Archeology of the Human Sciences* (Vintage Books, New York)
Gaspar J, Gould P, 1981 "The Cova da Beira: an applied structural analysis of
    agriculture and communication" in *Space and Time in Geography* Ed. A Pred
    (Gleerup, Lund, Sweden) pp 183-214
Gatrell A, 1981 "On the structure of urban social areas: explorations using q-analysis"
    Transactions of the Institute of British Geographers, new series 6 228-245
Goldblatt R, 1979 *Topoi: The Categorical Analysis of Logic* (North-Holland,
    Amsterdam)
Gorenflo L, 1982 "The mathematical structure of hunter-gatherer ecology I:
    Preceramic Oaxaca" in *Excavations at Cueva Blanca* Eds V Flannery, F Hole
    (Academic Press, New York) forthcoming
Gould P, 1976 "The languages of our investigations: Part I, Graphics and maps"
    InterMedia 4 13-16
Gould P, 1977a "The languages of our investigations: Part II, Algebras" *InterMedia*
    5 10-14
Gould P, 1977b "Learning how to steer: thoughts from Swedish and Nigerian
    comparisons" in *Geographic Humanism, Analysis and Social Action* Ed. D Deskins
    Michigan Geographical Publication 17, Department of Geography, University of
    Michigan, Ann Arbor, Mich., pp 377-390
Gould P, 1978 *How Shall We Classify Television Programs?* International Television
    Flows Project, Department of Geography, The Pennsylvania State University,
    University Park, Pa
Gould P, 1979a "Signals in the noise" in *Philosophy in Geography* Eds G Olsson,
    S Gale (Reidel, Dordrecht) pp 121-154
Gould P, 1979b "Geography 1957-77: the Augean period" *Annals of the Association
    of American Geographers* 69 141-151
Gould P, 1979c "Gender and society: an algebraic perspective from somewhere along
    the continuum" and "Pages from a conference notebook: some written, graphic and
    algebraic thoughts on gender and society" Meeting on *Gender and Society* February
    1979, Social Science Research Council, New York; available from the author at
    Department of Geography, The Pennsylvania State University, University Park, Pa
Gould P, 1979d "Mathematics and the social sciences: a perspective to $(1979+N)$"
    a report prepared for the Social Science Research Council, New York, December
    1979; available from the author as above

Gould P, 1979e *Dinâmica de Poliedros: Uma Introducão para Cientistas Sociais, Géografos e Planeadores* Estudos para o planeamento regional e urbano, number 9, Estudos de Geograficos, Lisbon

Gould P, 1980 "*Q*-analysis, or a language of structure: an introduction for social scientists, geographers, and planners" *International Journal of Man-Machine Studies* **13** 169-199

Gould P, 1981a "A structural language of relations" in *Future Trends in Geomathematics* Eds R Craig, M Labovitz (Pion, London) pp 281-313

Gould P, 1981b "Letting the data speak for themselves" *Annals of the Association of American Geographers* **71** 166-176

Gould P, Gatrell A, 1980 "A structural analysis of a game: the Liverpool v Manchester United Cup Final of 1977" *Social Networks* **2** 247-267

Gould P, Johnson J, 1980a "The content and structure of international television flows" *Communication* **5** 43-63

Gould P, Johnson J, 1980b "National television policy: monitoring structural complexity" *Futures* **12** 178-190

Gould P, Sugiura N, 1980 *One Day in the Life of Japanese Television* International Television Flows Project Report number 10, Department of Geography, The Pennsylvania State University, University Park, Pa

Graves R, 1966 *Robert Graves: Poems Selected by Himself* (Penguin Books, Harmondsworth, Middx)

Gregory D, 1978 *Ideology, Science and Human Geography* (St Martin's Press, New York)

Haack S, 1979 "Do we need fuzzy set theory?" *International Journal of Man-Machine Studies* **11** 437-445

Häbermas J, 1971 *Knowledge and Human Interests* (Beacon Press, Boston, Mass)

Häbermas J, 1973 *Theory and Practice* (Beacon Press, Boston, Mass)

Häbermas J, 1979 *Communication and the Evolution of Society* (Beacon Press, Boston, Mass)

Hägerstrand T, 1953 *Innovationsforloppet ur Korologisk Synpunkt* (Gleerup, Lund)

Hertz H, 1956 *The Principles of Mechanics Presented in a New Form* (Dover, New York)

Hilton R, 1969 *Algebraic Topology: An Introductory Course* (Courant Institute of Mathematical Sciences, New York)

Hofstadter D, 1979 *Gödel, Escher, and Bach: An Eternal Golden Braid* (Basic Books, New York)

Janik A, Toulmin S, 1973 *Wittgenstein's Vienna* (Simon and Schuster, New York)

Johnson J, 1975 *A Multidimensional Analysis of Road Traffic* Ph D thesis, University of Essex, Colchester, Essex

Johnson J, 1977 *A Study of Road Transport* Department of Mathematics, RR-11, University of Essex, Colchester, Essex

Johnson J, 1980a *Describing and Classifying Television Programmes: A Mathematical Summary* International Television Flows Project Report number 4, Department of Geography, Cambridge University, Cambridge

Johnson J, 1980b *Q-transmission in Simplicial Complexes* International Television Flows Project Report number 11, Department of Geography, Cambridge University, Cambridge

Jung C, 1961 *Memories, Dreams and Reflections* (Vintage Books, New York)

Jung C, 1964 *Man and his Symbols* (Doubleday, New York)

Kac M, Ulam S, 1968 *Mathematics and Logic: Retrospect and Prospects* (Praeger, New York)

Kessler S, McKenna W, 1978 *Gender: An Ethnomethodological Approach* (John Wiley, New York)

Laws J, 1975 "The psychology of tokenism" *Sex Roles* **1** 51-67
Leach E, 1969 *Genesis as Myth and Other Essays* (Jonathan Cape, London)
Lee D, 1979 *Plato: Timaeus and Critias* (Penguin Books, Harmondsworth, Middx)
Linhart J, 1973 "Uncertainty epidemics among interacting particles" *Il nuevo Cimento*
    13A 355-372
Magee B, 1978 *Men of Ideas* (The Viking Press, New York)
Marchand B, 1974 "Quantitative geography; revolution or counter-revolution?"
    Geoforum **17** 15-31
Marchand B, 1978 "A dialectical approach to geography" *Geographical Analysis* **10**
    105-119
Marchand B, 1979 "Dialectics and geography" in *Philosophy in Geography* Eds S Gale,
    G Olsson (Reidel, Dordrecht) pp 237-267
Marsh R, 1956 *Logic and Knowledge: Essays 1901-1950* (Allen and Unwin, London)
Martin-Löf P, 1969-70 *Statistika Modeller* (Institutet för Föräkringsmatematik och
    Matematisk Statistik, Stockholm)
Melville B, 1976 "Notes on the civil application of mathematics" *International Journal
    of Man-Machine Studies* **8** 501-515
Morrison P, 1957 "The overthrow of parity" *Scientific American* **196** 45-57
Mulhall D, 1977 "The representation of personal relationships: an automated system"
    International Journal of Man--Machine Studies **9** 315-335
Murray E, 1979 *Caught in the Web of Words* (Oxford University Press, Oxford)
Nicholis G, Prigogine I, 1977 *Self-Organization in Non-Equilibrium Systems* (John
    Wiley, New York)
Olsson G, 1980 *Birds in Egg: Eggs in Bird* (Pion, London)
Pears D, McGinnis B, 1961 *Wittgenstein's Tractatus Logico-Philosophicus* (Routledge
    and Kegan Paul, Henley-on-Thames, Oxon)
Poston T, Stewart I, 1978 *Catastrophe Theory and Its Applications* (Pitman, London)
Rosen C, 1976 *Schönberg* (Fontana, London)
Russell B, 1956 "Mathematical knowledge as based on a theory of types" in *Logic
    and Knowledge: Essays 1901-1950* Ed. R Marsh (Allen and Unwin, London)
    pp 59-102
Sayre A, 1975 *Rosalind Franklin and DNA* (W Norton, New York)
Seymor-Smith M, 1970 *Robert Graves* [The British Council (Longmans Group),
    London]
Simon H, 1979 *Models of Thought* (Yale University Press, New Haven, Conn.)
Simon H, 1980 *Items* **34** 6-7 (Social Science Research Council, New York)
Steiner G, 1970 *Poem into Poem* (Penguin Books, Harmondsworth, Middx)
Steiner G, 1978 *Has Truth a Future?* (BBC Publications, London)
Thom R, 1975 *Structural Stability and Morphogenesis: An Outline of a General
    Theory of Models* (Benjamin, Reading, Mass)
Wheeler J, Patton C, 1977 "Is physics legislated by cosmogony?" in *The Encyclopedia
    of Ignorance, Volume 1 The Physical Sciences* Eds R Duncan, M Weston Smith
    (Pergamon, London) pp 19-35
Whittaker E, 1960 *A History of the Theories of Aether and Electricity: Volume II,
    The Modern Theories 1900-1926* (Harper Torchbooks, New York)
Wilson A, 1974 *Urban and Regional Models in Geography and Planning* (John Wiley,
    Chichester, Sussex)
Wilson T, 1976 *Thirteen Perspectives on the Grammatical Landscape of Social
    Sciences* MS Thesis, The Pennsylvania State University, University Park, Pa
Zadeh L, 1975 *Fuzzy Sets and Their Applications to Cognitive and Decision
    Processes* (Academic Press, New York)

Philosophy in the construction of geographic reality

Helen Couclelis

"In our youth we looked more scientific".

> R D Specht (1964) commenting upon
> changes in Rand's approach to systems
> analysis since the 1950s.

In search of the promised land

Philosophizing in an empirical science is a sure sign of trouble[1]. Just as two decades ago practically anyone with a good mind and an adventurous spirit would 'go quantitative', philosophical speculation nowadays attracts a growing part of the academic élite in geography. To those who feel they may have reached the end of the road in pursuing the lines of current research traditions, the clouds of philosophy seem to promise a vantage point sufficiently lofty to allow the general layout of the disciplinary maze to be discerned.

The "search for common ground" alluded to in the title of this book reflects a major motive behind the growing concern with philosophy; namely, the hope for the discovery of some unifying perspective in a discipline plagued by excessive fragmentation of research and an increasing lack of mutual understanding among researchers. Yet the real significance of these problems is not immediately evident. Fragmentation of research is not necessarily in itself an evil: the world is after all a very varied place, and we need all the different perspectives we can get. I would rather see some blows exchanged than subscribe to any kind of dogmatism in the name of disciplinary unity, be it a dogmatism supported by some philosophy or not.

Nor is the unifying potential of philosophy something to be taken for granted. On the contrary, philosophy itself seems to foster the variety of perspectives that make up today's geography. As long as there are people around with different temperaments, beliefs, and ideologies, and as long as they have something to say, there should be room for positivist geography, for marxian geography, for Vienna-circle geography, for Frankfurt-school geography, for Lyon-group geography, and so on. The more, the merrier. Moreover, a lack of mutual understanding among researchers working in different areas of a particular field is by no means unique to geography. I once witnessed the efforts of a general relativist

[1] "It is, I think, particularly in periods of acknowledged crisis that scientists have turned to philosophical analysis as a device for unlocking the riddles in their field" (Kuhn, 1970, page 88).

trying to explain Riemannian curvature tensors to a low-temperature physicist, and I can assure you that it was not easy. The formation of closed, introverted groups of researchers mumbling to themselves in some esoteric lingo is a phenomenon which is even more pronounced in the 'hard' sciences than it is in geography. Although this tendency to overspecialization is enough cause of concern to epistemologists, there is no immediately obvious reason why it should worry geographers so much.

To my mind, what 'fragmented' geography lacks compared with 'fragmented' physics (or any other discipline not presently preoccupied with philosophical questions) is not so much a sense of unity as a *sense of direction*: clearly not that *linear* sense of direction conveyed by progression on a path trodden by one and all; such a conception would be consistent with a unidimensional view of reality, a most inappropriate view once we have named our kind of geography 'human'. In emphasizing the phrase "sense of direction", I use it in the same way an explorer of unknown territory might use it—as the knowledge of North and South, Here and There, Back and Forth. What we both need is some kind of map, plus a system of reference to tell us where we stand, how we fit in, how our work relates to that of others, how we can tell if we are making progress in the field of geography. Then, in a miraculous Gestalt switch, the simpletons playing with odd-shaped little pieces of coloured wood could suddenly be seen to be serious people working at different corners of the same jigsaw.

If the construction of such a map is indeed to be accepted as a central problem in today's geography, then the current concern with philosophy is amply justified. For in that case the pertinent questions have much less to do with geography as empirical subject matter than with geography as a body of methodological, epistemological, ideological, and ethical presuppositions in need of critical examination and logical analysis. What makes the consideration of these issues the geographer's rather than the philosopher's task is the recognition of the fact that they cannot be examined independently from the empirical base of geography. In particular, there is increasing evidence that the cognitive *content* of geographic theory, the part of reality represented in the theory, is strongly interrelated with its *form*—the theoretical language, and the logical, methodological, and ideological presuppositions involved in its expression.

In particular, there are two distinct critical approaches to the analysis of geographic theory which are directly relevant to this issue. One approach departs from ethical and ideological considerations inspired by moral, political, and social philosophy, and seeks to determine which part of reality *ought* to be represented in geographic theory, eventually reaching conclusions regarding the form (theoretical language) most appropriate for the expression of meaningful and relevant theory. The other approach departs from strictly rational considerations derived from

epistemology, mathematical and linguistic philosophy, and logic, and seeks
to determine which part of reality *is* actually being represented in a given
body of geographic theory, eventually reaching conclusions regarding the
limits of the possible cognitive content corresponding to each particular
theoretical form.

These two *critical* approaches, which may be called *normative* and
positive respectively, are to be contrasted with ordinary subject-matter
geography (whether quantitative or not), which is closely in touch with
empirical observation and is in no direct contact with philosophy
(figure 1). I think it is possible to identify roughly the above three
approaches with Habermas's three traditions of inquiry[2], although the
detailed interpretation of the present scheme could differ in important
respects from Habermas's. For example, by focusing on the logical, rather
than on the experiential, intuitive, and sympathetic dimensions of understanding (which in this case is *not* "verstehen"), the term *hermeneutic*
here takes on a more specialized sense than is normally the case in
phenomenology. Yet both etymologically ($\dot{\epsilon}\rho\mu\eta\nu\iota\alpha$ = [textual]
interpretation), and by virtue of accepted definitions (through the
determination of the frame of meaning within which a particular utterance
is significant), this use appears legitimate.

With due apologies to Habermas for the unlicensed adaptation, we may
go on to note a few thoughts suggested by the above scheme. The three

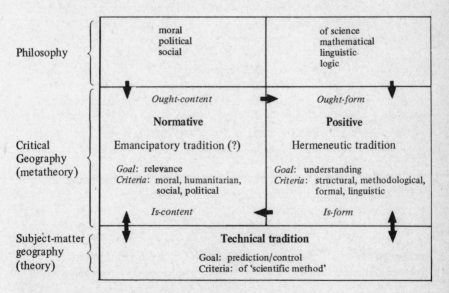

Figure 1. The traditions of geographic inquiry.

[2] The three traditions enunciated in his inaugural address at Frankfurt in 1965
(Habermas, 1971) summarized in Gould, "Is it necessary to choose?", (pp 76-78,
this volume).

approaches distinguished are not equivalent, nor are they totally ordered: hierarchically, they occupy between them one 'basic' or 'primary' level of statements about facts, and one 'critical' or 'meta' level of statements about statements about facts. This means that while in principle the technical approach can proceed indefinitely on its own resources by feeding on further and further empirical observations, the two critical approaches would soon dry up without the 'food for thought' provided by subject-matter (technical) geography. Thus, although technical geography without the insight provided by critical geography may be blind, critical geography is threatened with death by irrelevance if, instead of reflecting on the statements of technical geography, it wastes itself in self-indulgent introspection, and gets caught in the endless spirals of statements about statements about statements ... about facts. Clearly, technical geography must go on, and the more its vision is improved by the twin lenses of the hermeneutic and the emancipatory approaches, the less it will be necessary in future days to hold special seminars at which critical geography is treated as a separate and slightly superior species. My conception of a successful geography is of a discipline which does not turn to what philosophy is available seeking guidance from above, but is *inherently* philosophical as much as it is technical, eventually becoming itself a source of new material for philosophy in the way the most advanced branches of theoretical physics have done. And if some day we hear of a conference of professional philosophers entitled *Geography in Philosophy*, then we will know for sure that we are getting somewhere in our search for common ground.

What does an urban model really represent?

The ideas I wish to discuss in the following belong in the second tradition of critical inquiry, the hermeneutic tradition as defined in the preceding paragraphs. They concern the search for the 'frame of meaning' within which the class of spatial interaction models is logically valid[3]. This class is taken to include all the quantitative models in which urban or regional structure is expressed in the form of patterns of population movements across space, from the roughest social physics model to the most sophisticated microtheory, provided that the effect of some kind of *distance* is explicitly accounted for.

The research was triggered by a sense of wonder at the fact that such models would 'work' at all. As a newcomer to the trade, I was faced rather suddenly with all the assumptions, the simplifications, the choice of variables, the descriptions and the explanations of the urban system as represented by current spatial interaction models: the Featureless Plain, the Free Market, Perfect Information, the just-as-Perfect Competition,

[3] What follows is based on my Ph D dissertation entitled *Urban Development Models: Towards a General Theory* (Couclelis, 1977).

the Principle of Least Effort (this must have been before the days of mass jogging), egotistical Rational Man, apolitical Satisficing Woman, and so on. What bothered me in all this was not so much the simplification in itself: much rather, it was the fact that the urban system as represented in these models appeared as self-contained and self-sufficient, as if factors such as public housing policies, the system of taxation, the legislation of land ownership and transaction, the constraints on the use of private property, the sociopolitical system, history, culture, custom, and ideology had never played any role in city-shaping processes. All this went against the whole of my experience and intuition about how things actually happen in a city or region. Theoretical constructs ignoring these realities should never have achieved agreement with empirical data, and yet in a perverse way they did—in fact, the better the agreement the more naive they appeared. It was no use contesting the results of the statistical tests. A particular version of the Lowry model, I was told, had been used reasonably successfully in cities of the United States, in West Germany, repeatedly in Britain, in Jugoslavia, and in several developing countries of South America, Africa, and the Middle East. I was urged to find out whether it would work in the case of Athens just as well, but I saw no point in the exercise. I had by then little doubt that the Lowry model would have worked on Mars, if only the little green men would have stood up to be counted. The apparently universal applicability of such models, in defiance of every difference in sociopolitical, institutional, and cultural conditions observable from one context to the next, seemed paradoxical enough to prevent the outright dismissal of the modelling approach as irrelevant technocracy. In face of all the statistical evidence it was difficult not to accept that these constructs represented something —but what?

We are told that an urban model is an approximate representation or simplification or partial view of the urban system: but who has ever seen an 'urban system' to know what the approximation approximates, what the simplification simplifies, of what full view this is a part? Our direct experience is of cities and towns; *the* urban system, on the other hand, is an utterly obscure abstraction. What is its relation to the world of experience? What is its relation to the scores of existing urban models, all of which presumably represent aspects of it, yet no two of them give quite the same account of events and behaviour? Why is there often such a gulf between what models tell us of the behaviour of urban systems, and what we empirically know of city-shaping processes? Is the urban system something that could eventually be unambiguously defined, or must it forever remain elusive yet 'there', as inaccessible as a Platonic idea?

These questions about *the* urban system really concern the existence and nature of a *domain* of spatial interaction models, a logical frame of meaning within which these constructs 'make sense'. The tentative answer given in the study reported here (Couclelis, 1977), is that this domain exists as a coherent conceptual entity, that it can be defined as a theoretical *structure*, and that this structure can be obtained deductively, independently from experience, on the basis of the properties of the theoretical language used in the formulation of spatial interaction models. The structure can be seen as an intermediate abstraction interposed between whatever we may understand as the 'real world' on the one hand, and the spatial interaction models on the other. It bridges the logical gap existing between that which these models actually represent and corresponding aspects of the world of experience (figure 2). In addition to providing some unexpected insights into the meaning of spatial interaction models in relation to 'reality', this structure, though still to a large degree speculative, raises a considerable number of theoretical, methodological, and philosophical issues in its own right.

Subjective causality

The logical frame of meaning mentioned above is found to be a multilevel structure representing a hierarchy of urban systems, of which the simpler ones can be associated with certain well-known urban models. But rather than being a mere typology of models, the structure must be seen as an integrated whole—a single entity presenting many different aspects. More specifically, the *entire* structure can be said to represent the total urban system, each tier of the hierarchy corresponding to the image of the same part of the real world defined at a different level of abstraction. In systems terms, the hierarchy is one of types of system behaviour of increasing variety and complexity, all systems being models of the same original. Since system behaviour is usually associated with causal or

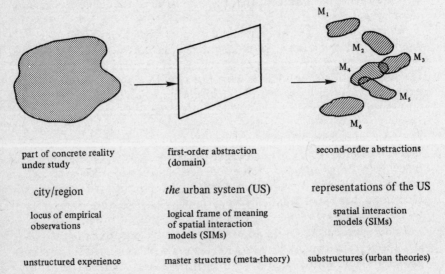

part of concrete reality under study	first-order abstraction (domain)	second-order abstractions
city/region	*the* urban system (US)	representations of the US
locus of empirical observations	logical frame of meaning of spatial interaction models (SIMs)	spatial interaction models (SIMs)
unstructured experience	master structure (meta-theory)	substructures (urban theories)

Figure 2. Urban reality and its models.

functional structures, it may be convenient at this stage to describe the hierarchy as a causal hierarchy, and the principle involved in its definition as *causal (dis) aggregation*, although it will be seen that the notion of causality is of no further consequence in the development of the structure.

The idea of causal disaggregation is based on the 'subjective' view of the cause-and-effect relation according to which causality is a conceptual frame necessary for our understanding of real-world phenomena, rather than some natural principle governing the phenomena themselves. This view stresses the relativity of causal constructs, and suggests that their validity is largely a function of the variables which the investigator chooses to include, since "the inclusion of further variables may alter the causal model almost completely" (Blalock, 1961, page 19). A further implication is that given two sets of events—events which appear causally connected—it is, in principle, always possible to intercalate additional causal links between them. Harvey (1969, page 392) illustrates such a procedure by which a two-variable cause-and-effect relation can be expanded into one with three, four, ... variables. A generalization of this hypothesis suggests that it should be possible, again in principle, to start from any given causal structure and to develop a sequence of sufficiently different causal models, each of which is derived from the preceding one, merely by the addition of further variables. This idea of causal *disaggregation* gives a first indication as to how a family of systems, ranging from broad statistical generalizations to elaborate causal explanations of the same phenomena, could possibly be hierarchically structured. The consideration of the inverse process of causal *aggregation* leads to further useful insights.

Consider a causal system represented in a state of arbitrarily detailed disaggregation, incorporating virtually all the variables that may have some effect on it, everything directly or indirectly affecting everything else. This can be thought of as a very large, very strongly connected network, the variables occupying the nodes, the causal arrows being the links. We may then imagine gradually removing classes of variables judged to be relatively 'less essential', together with their associated causal links. This will have the effect of creating holes in the original network, and if the process is carried on long enough the network will disintegrate into a small number of distinct elementary relations incorporating a few most 'basic' variables. Clearly the original system will be virtually unrecognisable at that stage, although the procedure of gradual abstraction apparently did nothing to distort it or alter it qualitatively.

The abstractive procedure must be carried out in some consistent way so that no essential causal links are severed while less essential ones are still operating, otherwise the remaining causal network will appear incoherent. Assuming that it is possible to distinguish relative degrees of importance within the class of variables operating in a system, we may speak of 'levels' of abstraction or thresholds beyond which certain classes of variables may be omitted as inessential. Raising the level of abstraction through the total causal structure will have the effect of filtering out further and further classes of relatively less essential variables, retaining only those that are more fundamental[4].

We may thus speak of slicing the total causal network at various levels of abstraction, each time ignoring all the variables which are secondary or inessential relative to a particular degree of generality. Each successive slice will present us with a simpler and more tractable image of the original system, although we will pay a penalty for the gain in clarity—an increasingly marked qualitative departure from the behaviour of the original.

The principle of causal aggregation, which can gradually lead us from a total causal network to some elementary statistical relations via a range of causal structures of varying complexity and behavioural type, provides us with the perspective needed to see the wide variety of existing urban theories and models in relation to each other and to the 'urban system' they purport to represent. All the theories and models of the class discussed here, from the simplest ones, which are also the better established, to the more elaborate ones, which are more interesting theoretically (but also more controversial), incorporate causal structures of varying degrees of complexity. Granted that all the models are roughly true (whatever that means) at their respective levels of generality, each of them may be seen as *a fragment of some total structure incorporating all the possible interactions between (urban) man and (urban) space*[5].

We may thus imagine the total structure of location–locator relationships, and consider what will happen if the principle of causal aggregation is applied. All that is needed is the possibility of distinguishing a discrete number of levels of abstraction, at each of which certain classes of variables may be eliminated as inessential. There are good reasons for considering seven such levels, although there is nothing particularly significant about that number.

[4] This idea is analogous to what is known as the "threshold effect" in cybernetics: "The existence of a threshold induces a state of affairs that can be regarded as a cutting of the whole into temporarily isolated subsystems ... [A drastic reduction of direct interactions] might be induced by a rising threshold. It will be seen that the reacting subsystems tend to grow smaller and smaller, the rising threshold having the effect, functionally, of cutting the whole network into smaller and smaller parts" (Ashby, 1956, page 66).

[5] The qualification *urban* before *man* and *space* indicates that we are already dealing with abstractions; that is, some model of man and some model of space compatible with spatial interaction models: we can never have real people and real (?) space in conceptual structures (Olsson, 1975). The significance of such a distinction will become evident in what follows.

'Simplifying' the urban system

The following sequence of levels of abstraction can be seen as a hierarchy of self-contained abstract urban systems which become simpler and simpler as further and further classes of variables are eliminated. The process should normally be expected to lead to some 'simplest' representation of an urban system, together with a set of 'most fundamental' underlying causal relationships. To some extent this is what occurs, although what emerges at the end belies any expectations of profound theoretical revelations.

Level 7. The starting point is the complete network of all possible people–space interactions. It is at this level that all the variables of the urban system operate simultaneously, every individual affecting and being affected by the total environment in a different way, with all the relationships changing constantly. In such a nebula of interactions no structure can be distinguished (or exists?), and thus nothing can be said of level 7 except that it contains all the causal structures that may conceivably be described. Nevertheless, from the empirical point of view this is the most significant level of all, since this is the one which corresponds to all the real and concrete decision environments in which all the real and concrete urban decisions are made. Furthermore, whatever we can observe, count, and measure in the urban system is necessarily the outcome of the behaviour of this complete, 'real' network—we cannot make measurements in abstract worlds. Each element of urban structure is as it is, and where it is, at a particular time as a result of a combination of circumstances involving an unknown number of major and minor decisions by an unknown number of actors, each one of them responding to his own personal perception of the situation. Thus the urban system of which we have direct perception is also the one of which we can have no objective knowledge.

Level 6. Given the impossibility of knowing the individual mental processes behind each city-shaping decision, we are obliged to objectivize and consider what 'anyone' would have done at a particular time given a simplified decision environment in which variables of a strictly personal nature—such as emotions, tastes, personal values, beliefs, hopes, expectations, states of mind, habits, misunderstandings, misconceptions, preconceptions, misinformations, anticipated changes, memories of past similar cases, and so on—play no role. At the same time we must eliminate as redundant whatever traits of the urban environment are only relevant given these 'personal' variables. This leaves us with a diminished, substantially simpler causal network which reproduces the expected behaviour of an abstract entity called 'anyone' in a logically possible urban world where locations have no social connotations, visual qualities, images enhanced by propaganda, and so on. We find that such an abstract urban system is in many ways not unlike the original one in its behaviour, *except* that it appears much easier to control: 'anyone' can

understand what the 'common good' is, and all planning constraints that
do not contradict any more basic tendencies are automatically obeyed.
Level 5. We now take a further step into abstraction and suppress all the
variables that refer to imposed constraints on the use of locations (planning
constraints, legal constraints, etc), since these may be changed any time,
and thus do not represent what we might call truly 'fundamental'
characteristics. At the same time, any human characteristics that refer to
a propensity to obey such constraints also become redundant and must
also be eliminated. Again, we are left with a causal network considerably
simpler than the one before, having lost several feedback loops and other
multiple connections. This new abstract urban system has much in
common with the previous one except that the locating entities (the
second-degree abstractions of human beings), are now free to pursue their
own private welfare according to their individual preferences to the best
of their material means. This is a ruthless world in which the optimum
macrostate (greatest aggregate satisfaction for the available material means)
is achieved when the individuals with the most means (the 'rich') have
maximized their own utility.
Level 4. Abstracting one step further, we now suppress the individual
combinations of preferences which cause variations of behaviour within
otherwise indistinguishable contexts, and at the same time produce all the
spatial characteristics which are the objects of discrimination in such
situations. In the resulting and substantially simpler abstraction, locators
are compulsive players in the big game of free market competition, each
one of them striving to comply with the rules which will safeguard market
equilibrium—albeit of a rather naive form compared to that of level 5.
By this stage the urban system has become sufficiently simple to be at
least partially and approximately representable in a spatial interaction
model. In fact, some of the most advanced urban models available, of
which the Cambridge Land-Use–Transport models are typical examples[6],
represent various aspects of the urban system at this particular level of
abstraction.
Level 3. We may simplify one step further, and suppress the variables
which bring about the market competition situation: in other words, all
the information about the differential bidding ability of the population,
and the corresponding information about the varying desirability or
intrinsic quality of locations. What is left is a highly idealised 'egalitarian'
world, where space is allocated to functional groups (activities) according
to functional needs in a way that is largely predetermined, so that the
locators are now hardly anything more than cogs in the mechanism of
ecological equilibrium. However, many traits of the previous abstract
system are still recognisable, although, given the nearly total lack of choice

[6] For example, Echenique (1977).

implied by the causal mechanism, it is difficult to argue that we are still dealing with decisions of rational beings—even very generally and in the aggregate. Because of the uncompromising simplicity of what remains of the causal network at this stage, the corresponding urban system (or at least a partial aspect of it) could be successfully represented in Ira Lowry's (1964) original "Model of Metropolis".

Level 2. We may still insist in going deeper, to uncover the even more general and fundamental features of the urban system. This time we suppress all the variables referring to urban activity and land use both from the urban population and from urban space, so that we cannot distinguish any longer between, say, residents and employees, nor between residential areas and employment centres. The causal system (or perhaps we should say *set* of causal systems, since by this stage the network has disintegrated into several isolated parts) emerges even more clearly as the members of the population appear to be driven by forces beyond their ken—the forces of *gravity*. Indeed, it is precisely at this stage that the various elementary formulations of the 'gravity law' are rediscovered [Clark's law, Stewart's laws, Wilson's family of basic spatial interaction models (Wilson, 1970a)], and obviously there can be no question of rational calculating behaviour, since all the human characteristics that could possibly be invoked to support such a hypothesis have long since been eliminated from the system.

Level 1. Intrigued by the discovery of such strong 'natural forces', we wish to find out what lies 'behind' the law of gravity, and so proceed to suppress one further double class of characteristics—those that describe the members of the population as bodies that can be counted and located in space, and those that describe urban space as a map on which such bodies can be located. What remains is the general idea of a population of material particles existing in a space, and the concomitant idea of a space in which material particles can exist. However vague this may be, there are still some traces persisting from the previous causal network, some very clear elementary relations describing the nature of that material space. It appears to be a space that is inhomogeneous and 'knotty', so that the population in it is more likely to be found concentrated in clusters which are denser at the centre than in any other form of distribution[7].

Level 0. There is now only one further step left to full abstraction as we proceed to suppress the last remaining empirical information, namely, the knowledge that we are dealing with a population of material objects, and

[7] This is one of the points most difficult to appreciate in a sketchy and verbal description of the model. The fundamental inhomogeneity of space at level 1 (and all subsequent levels) is demonstrated elsewhere in the study by means of a statistical argument based on Coleman's (1964) "method of residues". See Couclelis (1977).

a space which is a physical, rather than a mathematical space[8]. It would
appear that we have now reached the very foundations of the urban
system. Yet the 'ultimate truth' which we discover is of monumental
triviality: the featureless, infinite, isotropic space of pure Euclidean
geometry. This disappointing result deserves some further comments.

Things and attributes

The preceding discussion has shown that a process can be described by
which, starting with the complete urban system of all immediate individual
perceptions, we gradually suppress corresponding classes of human and
spatial variables until there is nothing left but a fully abstract and formal
framework of relationships between ideal mathematical objects, from
which even the faintest trace of human presence has disappeared. Equally
well we may say that by that stage all traces of urban space as we know
it have disappeared, since the E^3 continuum, with its uncountable infinity
of identical and dimensionless points, is strictly speaking not a space
consistent with countable, discrete, extensive material events (Atkin,
1974a, pages 82–104; Whitehead, 1920, page 23). Thus Euclidean space
appears as an ideal limit, a 'primitive' from the point of view of urban
theory as defined here (although by no means a primitive from the
mathematician's point of view).

The process of abstraction as described above poses some very puzzling
problems. As one of the soundest and most necessary principles of
rational knowledge, abstraction is normally expected to uncover the most
basic and fundamental features of the concrete system investigated. Yet
the qualitative changes from the original system that occur over a small
number of steps (changes as extreme as that from the urban world of
immediate perception to pure geometry) make it more correct to speak
of *different systems* rather than of various simplifications of the same
original. And this is brought about *merely by eliminating object
characteristics*: no explicit change in procedure, method, approach, or
philosophy is involved. This is intriguing for if there are real 'things'
(people and space) interacting in ways in which the passive observer (the
modeller) plays no role, one would normally expect the nature of such
interaction to remain basically the same, whether the outside observer
considers all the characteristics of these things or not. If, however, at
each step of the process of abstraction these things appear to be behaving
according to different laws, it may not be unreasonable to venture that
we are really dealing with different objects each time.

---

[8] The introductory remarks of Carnap to Reichenbach (1958) are pertinent here:
"It is necessary to distinguish between pure or mathematical geometry and physical
geometry. The statements of pure geometry hold logically, but they deal only with
abstract structures and say nothing about physical space. Physical geometry describes
the structure of physical space."

This viewpoint allows two conflicting hypotheses to be formulated:
(1) there exist things in time and space which have (known and unknown) characteristics; and
(2) the spatiotemporal clusters of known characteristics *are* the things. Both these views coexist side by side in modern science, and may be called the *atomic* and the *plenum* ontology respectively (Hooker, 1973). The first and more traditional view, which grants things an existence somehow independent from their properties, entails that object identity must persist throughout any process of abstraction, simplification, or generalization by which characteristics are suppressed. This is a basic tenet of empiricist thinking, and the whole philosophy of urban modelling is based on that assumption; namely, that the real urban system will continue behaving roughly 'like itself' in any simplified representation which maintains some of its more fundamental (whatever these may be) characteristics. This is because, according to this view (Hooker, 1973, page 211):

"... all laws (in particular, the laws of change) are reducible to fundamental laws concerning spatiotemporal relations among fundamental individuals [the 'things'], and the fundamental properties."

In contrast to this, the second view implies that each different cluster of characteristics is simply a different abstract object, and it cannot be expected to behave like another, sufficiently different, cluster of characteristics (another object) even if both are abstractions from the same 'real thing'. Thus (Hooker, 1973, page 211):

"The laws of nature will concern fundamentally the relations amongst properties and property-complexes and again, prior to any roles which they might play in actual scientific theories, all such relations are on equal footing."

This may be called the *combinatorial* view of scientific laws, according to which it is the various combinations of possible object characteristics that give rise to the laws, rather than any relations among fundamental individuals which happen to be carriers of such characteristics. Combinations of characteristics are thus *abstract objects* existing in their own right in suitable *abstract worlds* [9].

If this is so, and to the extent that urban models incorporate laws of the urban system, we must expect their validity to be dependent on the nature of the abstract objects (the clusters of characteristics) which they involve, rather than on the nature of the concrete entities from which such characteristics have been abstracted [10]. That this is a plausible

[9] For a bold expression of the 'plenum' ontology see also Whitehead (1920, page 21): "It is not the substance which is in space, but the attributes ... Space is not a relation between substances, but between attributes ... ."
[10] There is an interesting parallel here with Lancaster's work on consumer-goods characteristics (Lancaster, 1966).

interpretation can be seen from a cursory examination of the entity 'people' as it appears at the various levels of the abstraction process described above, and in the context of the overall behaviour of the urban system at these same levels. Starting at the level of the concrete world of experience, we have:

At level 7: fully conscious human beings making real decisions
At level 6: docile anthropoids fulfilling the planners' wishes
At level 5: the 'rational men' of classical economics behaving rationally
At level 4: game-playing automata playing the game of perfect market equilibrium
At level 3: cogs in an ecological clockwork
At level 2: material objects rolling along the paths of gravity
At level 1: a swarm of particles, the individual positions of which can only be given in probability terms
At level 0: void.

Similarly, we find that the entity 'space' also undergoes a series of transformations, although this is more difficult to appreciate without recourse to mathematical description. In the same process of abstraction, urban space is non-Euclidean down to the very last step. At least this is what emerges when the attempt is made to develop the structure deductively from first principles; that is, starting from the formal end of the hierarchy (level 0) and proceeding upwards by gradually adding further and further classes of characteristics until, ideally, all possibly relevant characteristics have been included. The process of gradual *abstraction* from the actually perceived to pure formalism has the advantage of being relatively easy to follow intuitively, although it also has the disadvantage of being impossible to carry out in practice. This is because the starting point, the complete real-world system of all possible interactions, is, almost by definition, impossible to represent. On the other hand, by following the reverse procedure we may never actually manage to reconstruct the total net of relations at level 7, although we are in fact retracing the elementary logical steps implicit in the formulation of spatial interaction models, gradually rediscovering their basic mathematical forms as substructures of a more general or 'meta' structure.

Central to the approach is the treatment of the two basic interacting entities, urban man and urban space, as *sets of characteristics* in accordance with what was said earlier about the plenum hypothesis in science. It should be clear that this premise implies no particular metaphysical view on the essence either of man or of space: it is merely an epistemological assumption found compatible with the kind of 'laws' discussed here. However, what it does imply is a particular kind of *model* for man and space, and this must now be discussed in some detail.

People and attributes

Several different conceptions of 'people' are used in the various branches of urban studies and geography, each one of them a *stereotype* describable as a function of a limited number of human characteristics or variables which are considered the most relevant for the particular purpose of each study. For example, we have:

(1) National statisticians' Man, who is countable, has age, gender, belongs to a socioeconomic class ...;

(2) Traffic engineer's Man, who is countable, makes trips, has or has not choice of transport ...;

(3) Economist's Man (Rational Man), who has income, needs, is intelligent, well informed, selfish ...;

(4) Politician's Man (Compromising Man), who has demands, social class, values, is misinformed ...;

(5) Urban psychologist's Man, who has habits, tastes, culture, sensitivity, limited understanding ...;

(6) Town planner's Man, who has place of work, of residence, shopping habits, political beliefs ...;

and so on.

Some of these abstractions have little to do with each other [for example, (2) and (4)], others are almost mutually incompatible [perhaps (3) and (5)?], yet there is some truth in all of them, since they all represent different aspects or manifestations or *roles* of the same entity—the whole, concrete individual. Because of the inextricable interrelationships among all these aspects or roles, the problem of their independent analysis has not been resolved. On the other hand, attempts at crossbreeding to produce the Man Of Bounded Rationality by combining (3) and (5), or the Man Of Urban Models, by combining (1), (2), and (3), create new difficulties, because the complexity and possible internal contradiction of the resulting profiles often impair their usefulness as analytical abstractions.

A truly comprehensive theory of urban systems must allow for all these aspects to coexist without undue complication and especially without contradiction. More generally, it must allow for every possible combination of characteristics that may be needed to describe or justify the behaviour of every possible (major or minor) urban decisionmaker at any particular moment in time. It is evident that no analytical stereotype of Urban Man, however sophisticated, can offer such flexibility.

The alternative is to define Urban Man not as a standard set, however large, of *preselected* characteristics, but as the set H of *all possible* human characteristics or attributes that may be relevant in the explanation of some aspect of his urban behaviour, at any particular time. In other words

$H = \{h: \ h$ is a human attribute of potential relevance to some aspect of the urban system$\}$.

We shall assume that this set is *finite* though arbitrarily large.

Once the set H has been thus defined, any particular role or *profile* associated with any person n at any time can be represented uniquely and unambiguously as a *binary code* in a rectangular array, matrix, or tableau, thus:

Time = t — Attributes

Individuals	h_1	h_2	h_3	...	h_i	h_j	...	h_r	h_s
n_1	1	1	0	...	0	1	...	0	0
n_2	1	0	1	...	1	1	...	1	0
$\vdots$									
n_w	0	1	1	...	1	0	...	0	1

where h_i is any relevant attribute—for example, 'is countable', 'is travelling to work', 'is in income group B'. It can be easily seen that all the analytical stereotypes mentioned earlier are among the profiles definable in the tableau, since each of them can be expressed as a standard string of the form

$$(h_1, \overline{h}_2, \overline{h}_3, ..., \overline{h}_i, h_j, ..., h_r, h_s) \ ,$$

(where $\overline{h}_n$ means that the nth attribute is absent), or, more simply, as the set of attributes

$$\{h_1, ..., h_j, ..., h_r, h_s\}$$

There are two obvious difficulties with such a representation. The first is that whereas any profile can be recorded on the incidence matrix as a binary string, not every such string represents a valid profile. For example, if h_{m} is the attribute 'is male', and h_{f} is the attribute 'is female', any code that contains

$$\frac{... h_{\mathrm{f}} ... h_{\mathrm{m}} ...}{... 1 ... 1 \ ...} \quad \text{or} \quad \frac{... h_{\mathrm{f}} ... h_{\mathrm{m}} ...}{... 0 ... 0 \ ...}$$

may be considered invalid. Similarly, no person can be at the same time an infant and an adult, or, more subtly, an infant and an industrial worker; no one can be 'at home' and 'on his way to work' simultaneously, and so on. Thus, the first problem is how to recognise invalid codes, or, better still, how to ensure that they do not arise. The second problem is the practical one of having to deal with binary strings that may be indefinitely long.

As regards the second of these difficulties, we notice that in urban theory it is not the individual members of the population that are of interest, but the different *roles* these members assume. This means that the relevant 'objects' at each moment in time can be exhaustively described by relatively short chains of attributes suitably selected from the universe H. Obviously many members of the population can play the same role simultaneously, and in fact most of urban theory as developed to date centres on roles that are very widely shared. This means that in most cases we shall be dealing with a small number of role groups, rather than with millions of individual people. However, it is important to stress that *in principle* the same procedure that identifies roles will also identify individuals if carried out a sufficiently large number of steps.

The solution to the first problem, that of generating valid profiles at will, can be reached by exploiting the *redundancy* of the set H of Urban Man attributes. In other words, we can employ the fact that once a particular attribute has been established in a profile a great many others are automatically eliminated from further consideration, whereas others become much more likely than before. An analysis of this property leads to the establishment of a partial order on the set H, which can be mapped into a *hierarchy of cover sets* of attributes[11]. For various reasons too involved to explain here this is taken to be a hierarchy of seven levels $(H_1, H_2, ..., H_7)$, each of them corresponding to one of the subsets of attributes $H_1, H_2, ..., H_7$ partitioning the universe H of

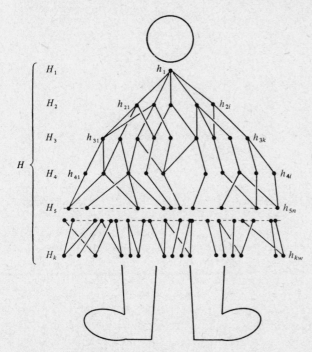

Figure 3. Urban Man.

[11] Given two sets Y' and Y, we say that Y', where $Y' = (... Y_i' ...)$, is a *cover* of Y if Y_i' are subsets of Y, and $\cup_i Y_i' = Y$. This hierarchical structuring of the set of Urban Man attributes is consistent with Russell's Theory of Types (Russell, 1908; Ayer, 1972, page 49), according to which propositions about objects, and in particular predicative statements, are arranged in a hierarchy.

all potentially relevant Urban Man attributes. These subsets are defined in the following manner:

$H_1 = \{h_1 : h_1$ is the attribute 'is a member of the population in urban region A'$\}$. This is a singleton set, corresponding to the universal upper bound of the ordered graph;

$H_2 = \{h_2 : h_{2i}$ is the attribute 'is in area (zone, cell, ...) b_i'$\}$, where b are all the parts of a given partition B of region A;

$H_3 = \{h_3 : h_{3j}$ is the attribute 'is engaged in process c_j'$\}$, where c are the elements of the set C of all space-related activities of interest to urban theory;

and so on to H_7.

Given the above partition, it can be shown that any path through the hierarchy $(h_1, h_{2i}, h_{3k}, ..., h_{7j})$ is a valid profile (figure 3).

To revert to the earlier representation of profiles as binary codes, the partial order representing Urban Man may be written as an ordered array of the form:

$$\left|\begin{array}{c|ccc|ccc|c|ccc} H_1 & & H_2 & & & H_3 & & & & H_n & \\ \hline h_1 & h_{21} & ... \ h_{2r} & ... \ h_{2i} & h_{3i} & ... \ h_{3s} & ... \ h_{3j} & ... & h_{n1} & ... \ h_{nt} & ... \ h_{nu} \ ... \end{array}\right|$$

in which case the profile $(h_1, h_{2r}, h_{3s}, ..., h_{nt}, h_{nu})$ corresponds to the code

$$\left|\begin{array}{c|ccc|ccc|c|ccc} 1 & & ... \ \ ... \ 1 & ... \ ... & ... & ... \ 1 & & ... \ ... & ... & ... & ... \ 1 & ... \ 1 \ \ ... \end{array}\right|$$

the empty spaces being all 0.

The attribute sheet

Individual codes can be tabulated in an accounting framework called the *attribute sheet*. The attribute sheet registers the characteristics of the urban population as only God could have known them. The supernatural power required to set it up is the ability to know the value (1 or 0) of all the relations between each possible attribute and every member of the population with perfect certainty, and this at each moment in time. A complete attribute sheet would give an exhaustive *description* (not explanation) of the instantaneous state of the urban system: where the members of the population are, what activities they are engaged in, what economic transactions arise through their activities, what material needs are satisfied through such transactions, what institutional constraints are obeyed in the expression of such needs, and what the personal reasons and motives are behind all the decisions and actions. Thus, the attribute sheet is an ideal model, not a device that could be used in practice. However, there is no great difficulty in imagining it as the remote limit of increasingly accurate and comprehensive tabulations of observed data.

The complete attribute sheet consists of seven sections, corresponding to the seven parts of the set H of human attributes as distinguished above.

It looks like the following:

Time = t	H_1	H_2	H_3	H_4	H_5	H_6	H_7
	h_1	$h_{21} \dots h_{2i}$	$h_{31} \dots h_{3j}$	$h_{41} \dots h_{4e}$	$h_{51} \dots h_{5m}$	$h_{61} \dots h_{6n}$	$h_{71} \dots h_{7w}$
n_1	1	0 ... 1 ... 0	0 ... 0 ... 1	1 ... 0 ... 0	0 ... 1 ... 0	1 ... 0 ... 1	0 ... 0 ... 1
n_2	1	0 ... 0 ... 1	1 ... 0 ... 0	1 ... 0 ... 0	0 ... 1 ... 1	1 ... 0 ... 1	1 ... 1 ... 0
$\vdots$							
n_w	1	0 ... 0 ... 1	0 ... 1 ... 0	0 ... 1 ... 0	0 ... 0 ... 1	0 ... 0 ... 1	1 ... 1 ... 0

The binary string that extends from H_1 to the last element of section H_7 may be considered, for all intents and purposes, to represent each member of the population uniquely, but codes that reach only to the end of H_6, or H_5, or H_4, etc, all represent perfectly *coherent profiles* of urban roles that are very probably (for the longer codes) or almost certainly (for the shorter) the same for several members of the population. It follows that valid attribute sheets can be set up with only 6, 5, 4, etc sections. Each of these would give a fully coherent, though incomplete, description of the instant state of the urban system.

For each of the seven possible code lengths, the exact number of similar profiles at each moment can be obtained by simple inspection. This number will, of course, represent the size of the group of people who are at that particular moment, and relative to the urban role represented by each profile, indistinguishable from each other. It is obvious that these groups become fewer and larger as information is lost; that is, as the codes become shorter. At the limit, when the code is restricted to H_1, there is only one group, the entire population. On the other hand, at the other extreme, when all possible information has been made available, there are no groups at all, only distinct individuals.

Urban time as a function of information
Each attribute sheet is only an instant picture. The relations we are interested in are of the kind that change with time, and an attribute sheet collected some time after a first one will show a different picture. It is possible to follow change as closely as desired by filling in attribute sheets at appropriate time intervals. A time series of attribute sheets will give an exhaustive description of the system's dynamics for that period.

The consideration of what may be an appropriate time interval leads to some interesting insights. Whatever the metaphysical essence of time may be, in the present context its flow can only be detected through the *changes* observed from one attribute sheet in a time series to the next: given a series that consists of identical sheets, it is impossible to say whether these are simultaneous replications of the same instant state of a dynamic system, or whether they represent the permanent configuration of a system that 'froze' in one of its states. In this sense, time is a change on a code pattern.

Purely combinatorial considerations show that time intervals measured as equal on ordinary clocks have a different value for the urban system as recorded on attribute sheets, according to the number of sections (1, 2, ..., 7) used in the representation—that is, according to the amount of information considered. This is because each further section that is added to an incomplete attribute sheet greatly increases the variety of distinguishable states and, therefore, the probability of observing some change somewhere in the system. In other words, at each step we introduce a greatly increased probability for change to occur between two consecutive sheets: *merely by adding information, we precipitate change*. This means that the attribute sheets, and consequently the more or less simplified abstractions of the urban system which they represent, become increasingly sensitive to time as further classes H_n of attributes are added. Such a property also extends to a purely spatial (as opposed to purely behavioural) account of the urban system, as well as to the results of the interaction between these two different views. In each case, it can be said that as more information is added, time seems to flow faster. We are thus able to define 'urban time' as a function of information (see also Atkin, 1978).

A graphic illustration can be given of the relative value of time intervals according to the length of the attribute sheet (figure 4). Since the shortest noticeable time interval is that for which at least one change is recorded, it is evident that the average absolute duration of the unit time intervals will contract as the probability for some change to be recorded increases: by the time one change has been observed on a (H_1) sheet, there may have been dozens on a (H_1, H_2) sheet, hundreds on a (H_1, H_2, H_3) sheet, and so on. At the limit, for the complete sheet, the duration of the unit time intervals will tend to zero, and time will appear to be *continuous*.

The increased sensitivity of information-rich systems to what we might call absolute time, can be related to the possibility of modelling highly simplified urban systems statically, as opposed to the necessity for dynamic modelling when dealing with richer abstractions. Associated

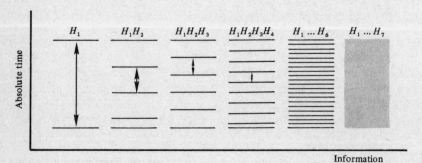

Figure 4. Relative urban time.

with the above considerations is also the idea of temporal *cycles*. It is obvious that a sheet of finite length can only give rise to a finite number of different configurations, and the shorter the sheet, the smaller their total numbers will be. Thus if we go on collecting attribute sheets for a sufficiently long period, sooner or later we will start getting arrangements already observed. If there is a periodicity in their recurrence (for whatever unknown reason), it is natural to isolate one cycle and study change within it; this will give an account of the system's dynamic equilibrium. It is evident that the *a priori* probability for such equilibrium to exist is much higher for the simpler systems, described by the shorter sheets where the total number of possible states is much smaller, than for the systems incorporating a larger amount of information. The same conclusion may be reached more formally, and it has far-reaching consequences as regards the *predictability* of the urban system's behaviour.

Urban space as a function of information

We shall now turn briefly to the concept of urban space. The concept of *space* in urban studies is the object of a confusion very similar to that surrounding the concept of Urban Man. The standard view of urban space in quantitative studies tends to be that of the isotropic, homogeneous three-dimensional continuum of Euclidean geometry. On the other hand, there are numerous claims (and proofs!) that urban space *is* non-Euclidean, or that there exist distinct entities like 'socioeconomic space', 'socio-political space', 'personal space', and so on, the geometries of which are non-Euclidean[12]. There is also, of course, the fully subjective, intuitive notion of urban space carried within each individual, a notion which appears almost entirely unanalysable even though it is probably the only real background to the decisions of the real (that is, concrete) individual. As with the many faces of Urban Man, the fact that all these conflicting conceptions of Urban Space exist in parallel, and are fruitful under certain conditions, must be taken into account in a truly comprehensive representation of the urban system.

In close analogy to what was said earlier about Urban Man, the view advanced here is that:

1. There is no such thing as an absolute urban space, independent of the *information* available for its description.

2. The various kinds of 'spaces' identified in different studies are only different *levels* along a hierarchy ranging from the entirely objective, universal, immutable space of Euclidean geometry to the entirely subjective, personal space perceived by each different human individual at each particular point in time.

---

[12] See, for example, Atkin (1974a), Harvey (1969; 1973, chapter 1), Lynch (1960), Massey (1974); "... each form of social activity defines its space; there is no evidence that these spaces are Euclidean or even that they are remotely similar to each other" (Harvey, 1969, page 53).

The concept of urban space as developed in the study allows for the whole of that spectrum to be incorporated in urban theory, each different level being defined as a function of available information. In particular, it is possible to describe a 7-stage procedure of defining urban space, starting with minimal information and gradually incorporating the entire set S of potentially relevant spatial variables. This is, therefore, a *cumulative* process, whereby at each step a further increment of information S_n (with $n = 1, ..., 7, \cup S_n = S$) is added to what was already known. The result is a sequence of qualitatively different spaces, none of which is Euclidean in spite of the fact that E^3 (the Euclidean three-dimensional continuum) is adopted as the starting point.

The initial step (step 0) consists of choosing Euclidean geometry as a starting point for the description of urban space. Then the first increment of information (S_1) is brought in whereby some 'primitive' geographical terms such as *location, distance, boundary, area* are substituted for the corresponding theoretical terms in the formal geometry. Because of the empirical content (information) implicit in the geographical terms, the result of this substitution (step 1) is *Space 1*, an interpreted formalism which differs from the original Euclidean geometry in certain very important respects. The fundamental modification to the original Euclidean framework is brought about by the substitution of *location* for *point*. In essence, a location is defined as a point with extension[13]. This operation transforms the Euclidean continuum into something qualitatively different, a space which is both *discrete* and intrinsically *inexact*. We call this first abstract space *Geographic-space*.

Next (step 2), we incorporate into Geographic-space the second increment of information (S_2), which consists of a system of coordinates (a reference grid), and a system of measurement, including length and area units within a certain range of physical magnitude; for example, m, km, km^2; or yard, mile, mile^2. The result is *Space 2*, called *Standard-space*, in which it is possible to identify and distinguish locations by determining both their absolute and relative positions. This allows us to map the relevant locations of urban space (those associated with urban 'events' of any kind) into a two-dimensional pattern characterised by relative size, shape, connectivity, grain, and so on. A three-dimensional map, registering height differentials in addition to the horizontal patterns, could also be obtained.

[13] The notion of a *point with extension* has been introduced in connection with various theories of *discrete space*. In Grünbaum's theory, a space is defined in which every region is composed of discrete spatial chunks, atoms or grains or quanta which have no spatial parts, though they are extended (see Nerlich, 1976, pages 156ff.). Whitehead (1920, page 86) uses a similar concept when he speaks of "event particles" as the ideal minimum spatial limits to events, to be contrasted with the extensionless points of pure geometry.

In fact, standard space comprises all the information needed for drawing something like a topographical map of the urban system.

Then the third increment of spatial information (S_3) is added to Standard-space (step 3). This information consists of the definitions and empirical information needed to distinguish various kinds of locations (and larger areas) corresponding to different urban activities, and to identify these with the various spatial elements of towns and cities—for example, the residential, industrial, service buildings and areas; the open spaces; the roads; and so on. The result is *Space 3* or *Adapted-space*, in which it is possible to describe urban space, location by location, as a particular arrangement of land uses and transport and service networks. This can be seen as 'colouring' the topographical map of Space 2 to convert it into a land-use map.

The same procedure is continued another four steps to Space 7. Spaces 4, 5, 6, and 7 each build on the preceding ones through a fresh import of information until, by Space 7, all possibly relevant information about urban space has been included. After Space 3, where the concept of land use first becomes meaningful, and as urban space acquires further and further classes of properties, it becomes gradually possible to define within it notions such as market value (Space 4), utility (Space 5), institutional and social determinants of land development (Space 6), and subjective valuations of urban space (Space 7). However, by the time we think we have reached Space 7, the quasi-formal description of the first step has degenerated into something so dependent on the observer-descriptor's personal perception that there may well be as many different accounts of the same part of urban space as there are observers to describe it at each moment in time.

Thus a multistage representation of urban space is possible over which the strictly objective, universal, timeless space of Euclidean geometry (the assumed 'minimal' description) gradually becomes the fully subjective, personal space experienced by each particular individual at each point in time (the 'maximal' description). The transition from each stage to the next is effectuated through an enrichment of the conceptual apparatus available for the description of urban space, so allowing the discrimination of additional differences within spatial categories which appeared homogeneous at a prior stage. In fact, the procedure is almost exactly parallel to that allowing the gradual description of Urban Man, and most of the comments which applied there can be transferred here: the gradual appearance, strengthening, and disappearance of *pattern* as more information is added; the increasing significance of *time* as we progress from formalism to concrete reality; etc. The one substantial difference in the description of Urban Space is the assumed existence of a complete, self-contained, and formal framework (Euclidean geometry) as the starting point.

The urban system

What we have achieved so far is the construction of a 7-level cumulative hierarchy of Urban Man *descriptions*, and a 7-level cumulative hierarchy of urban space *descriptions*. All the descriptions appear perfectly coherent and complete at their respective levels of generality, but nothing can ever be *explained* or predicted: in the case of urban space, all the spatial schemes encountered, from the fundamental inhomogeneity of Space 1 to the multitude of mental images carried by Space-7 observers, appear to be chance configurations. Similarly, in the case of the codes representing Urban Man, some arrangements may occur very frequently, others not at all; some may almost invariably follow, or precede, others; there may or there may not be cycles in a time series; and so on. It seems quite impossible to explain any regularities in the observed patterns.

This state of ignorance concerning the theoretical properties both of Urban Man and of Urban Space is overcome as soon as the synthesis of these two separate components of the urban system is attempted. Neither Urban Man nor Urban Space are determinate independently from each other, but the entire urban system, the result of their interaction, is found to be determinate to a large degree. As it turns out, the properties

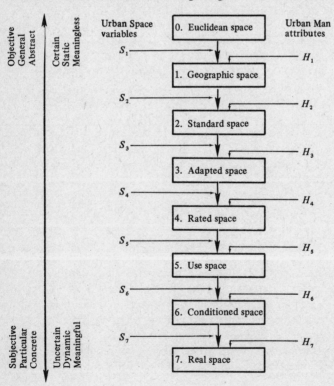

Figure 5. The urban system.

of Urban Space become logically predictable as a function of the properties of Urban Man. Conversely, the unexplained behaviour of Urban Man as reflected in the attribute sheet gradually becomes predictable as more information about Urban Space becomes available. The principle of the synthesis in question can be represented in a very simple diagram (figure 5).

The result of the synthesis is a cumulative hierarchy of internally coherent abstract worlds, all of which are occupied by 'urban systems' of varying complexity. Each of these abstract worlds is characterised firstly by a *field*, which incorporates all the constraints set by the information available in each case to the range of possible system states, and secondly by *laws*, expressing the most 'natural' distribution, against the biased background of the field, of otherwise unconstrained (equiprobable) events. Such laws happen to be structurally identical with the basic mathematical forms of current *spatial interaction models*.

Although all the details of the construction of the hierarchy cannot be given in this short essay, it is relatively easy to explain the basic lines of the principle involved. It is a question of making at any time the best possible use of a given amount of information without assuming anything that is not contained in that information. This is, of course, the whole idea behind the entropy maximising approach, now widely used in urban modelling, but what is different here is that the information is gained gradually and systematically in such a way that the most likely statement possible at each step is an elaboration on the most likely statement possible at the previous step[14]. A geometrical interpretation of the same principle can also be given; the information available at each stage can be structured into a geometrical space representing the combined result of all the known constraints which affect the probability of any event of a particular kind happening at a particular place. For the sake of illustration, we may think of a two-dimensional space which is distorted under the effect of different weights (probabilities) at each one of its points (locations). Such 'curved' two-dimensional spaces can be visualised in three dimensions as a series of hilly landscapes of increasingly complex topography, the first one being the perfectly flat surface of the Euclidean plane, perfectly free of (empirical) constraints in a world where there are no (empirical) events to be constrained. Since at each stage all the available knowledge is incorporated in the geometry of the corresponding space, total uncertainty exists relative to anything happening in that space external to the geometry, such as the event of a particular object ending up at a particular location. Over and above the bias represented by the uneven shape of the

[14] There is an obvious affinity between this idea and the Bayesian viewpoint of prior and posterior probabilities (see Batty and March, 1976), although the technical aspects of this possible connection have not been explored. The entropy maximising approach mentioned here is that developed by A G Wilson in a long series of publications (Wilson, 1970b; 1974).

space, we can assume no more than random influences affecting individual events and random distributions of events. Thus the least prejudiced statements possible at each stage can only be estimates of most probable distributions of objects derived on the basis of statistical-mechanical considerations, set against the 'loaded' background of the constraint space. These probabilistic statements, resulting from a combination of a state of analytical certainty on the one hand and encoded factual ignorance on the other, will be the *laws* of the corresponding abstract urban system. It is quite obvious that these laws will reflect the state of knowledge compatible with a particular level of abstraction, rather than any properties of the real-world system at the far end of the hierarchy, the end which is never reached.

In short, the combination of certain formal geometrical rules with a suitable selection of constraints (in the form of information) defines a probabilistic space or abstract world characterized by some *intrinsic structure*. Such a structure is fairly well defined at the early stages of the hierarchy, but becomes gradually more uncertain as the variety of the information introduced in the system increases, until by level 7 it seems to dissolve altogether. Of this sequence of structures the earlier, firmer forms are partially expressed in the family of spatial interaction models, the simpler forms of which are the more securely established. We thus reach two seemingly contradictory conclusions concerning such models: on the one hand, their truth is in a way *a priori*; that is, their basic forms are valid almost by logical necessity. On the other hand, they are entirely contingent on a particular choice of rules and information; their truth with respect to the 'reality-out-there' appears very relative indeed.

It may be noted that the particular choice of rules and information on which the validity of these models depends constitutes the body of *assumptions* associated with the corresponding theories. We may thus view the old problem of assumptions, 'realistic' and 'unrealistic', in a new light. A theory based on a set of blatantly unrealistic assumptions may be valid if these propositions happen to be descriptive of some particular abstract world in the hierarchy, provided that the theory itself represents part of the intrinsic structure of the space at that same level[15]. There are cases, however, when assumptions, or rather, combinations of assumptions can be genuinely unrealistic. This happens typically when they correspond to propositions which, given the logical relations holding

---

[15] Examples of level-defining assumptions are: isotropic plane—level 2; basic industry (or any other activity) as the functionally prior land use—level 3; perfect land market—level 4; utility maximising economic behaviour—level 4 or 5 (depending on the variables considered); etc. A further use of assumptions is to facilitate the description of levels already too complex for exhaustive treatment: in many cases aggregating, leaving out detail, fixing certain parameters, and so on, are level-internal operations that are *simplifying* in the ordinary sense—but simplifying a particular level, *not* 'the' urban system itself.

at the relevant level of the hierarchy, are mutually inconsistent. For instance, Angel and Hyman (1971; 1976) have shown by the method of geometrical map transformation that each of the two combinations: (1) the Euclidean plane and uniform transport facility, and (2) uniform population densities and uniform transport facility, is inconsistent. The same conclusion emerges as an implication of the present theory when it is shown that the laws characterising the relevant levels preclude, in the general case, the coexistence of conditions as above. Thus the distinction that must be drawn is really one between *empirically* and *logically* unrealistic assumptions, the former being a necessity in theories, and the latter a fatal flaw.

This novel interpretation of the nature and function of assumptions in theory building is just one instance of the more general insights that can be gained through an analysis of the properties of the general structure. However, it is not possible to go very far in any analytical sense as long as the description of the structure remains largely verbal—and hence highly speculative. The question of the choice of an adequate formalism poses itself because it is quite clear that conventional statistical and analytical methods, although sufficient to a certain extent for the exploration of individual levels in the hierarchy, are inadequate for the representation of the properties and relations characterising the structure as a whole.

The most concrete promise for a successful formalisation of the general structure is to be found in the work of Atkin (1972; 1974a; 1974b; 1974c; 1975), with which the approach described in this paper presents certain marked parallels. Atkin's work (known generally as *Q*-analysis or polyhedral dynamics) is based on algebraic topology and develops a combinatorial language for studying the *structure* of the relations existing between finite data sets. It appears that many of the basic ideas underlying the general model mentioned here can be given an interpretation in terms of the geometric concepts from which Atkin's mathematical language of structure is developed. Atkin's work is becoming familiar to geographers through recent writings of Peter Gould (1979a; 1979b; 1980), who recognised its great potential as a language capable of expressing the concerns of a new geography.

The structure and the map

The conceptual structure described in the preceding pages is just one instance of a 'map' for geography covering an important, though limited, area of theoretical concerns in the field. With the help of this map researchers can identify their respective positions relative to each other and to the whole of the area covered by the structure: the 'social physics' people at levels 2 and 3, the free-market people at level 4, the utility-theory people at level 5, the 'relations of production' people at level 6, the behavioural geographers (at least some of them) possibly at level 7.

Macro and micro, statistical and causal, static and dynamic, quantitative and qualitative—a large number of 'polar opposites' of current geographical research are synthesised and integrated in a coherent picture. There will be other such maps I am sure, and I hope that they will be more explicit than the present one with respect to the higher, far more interesting levels which are closer to our practical and humanistic concerns. While the validity of this particular framework remains to be proved, I hope that I have succeeded in drawing attention to the great potential interest of searching for the *frame of meaning* behind a given category of theoretical statements.

There should be no misunderstanding as to the status of the conceptual structure described in the preceding pages: it purports to represent nothing more than the 'domain' of a particular class of theoretical constructs, the spatial interaction models. It merely delineates the limits of their possibilities and their logical validity. Whether this is also an appropriate frame of meaning for the treatment of the kind of issues we may consider most relevant in today's geography is a question that cannot be asked within this particular mode of inquiry: this is where the complementary 'normative' or 'emancipatory' approach must take over. However, a number of other questions can at least be tackled within the present framework.

First, and supposing that the model is indeed valid, what do we make of that knowledge? In a way, the structure can be said to represent a 'geography of possible worlds'; it tells us that the fundamental laws expressed in the simpler spatial interaction models, being analytical, are inescapably true—but true in their respective abstract realms, not in the one real world in which we are interested. However disappointing (or rejoicing) this conclusion may be, these 'laws' could well be the only bits of definitive (that is to say, certain) knowledge we have in geography: we cannot afford to throw them away. There must be ways of using them intelligently both in theory building and theory application, but these have yet to be found. A lot of high-level basic research will probably be required, research into fundamental logical, mathematical, and linguistic structures rather than yet more empirical tests of incrementally embellished statistical models.

Second, if, as suggested by the present theory, there can be no laws of spatial interaction at level 7, which represents the actual world of experience, what other possibilities of systematic knowledge are left for the theoretical geographer to look for? The basic properties of the general structure suggest that the theoretical possibility for *prediction* diminishes rapidly as we move from the quasi-formal intrinsic structure of level 1 towards the information-rich end of the hierarchy until, at the limit, it seems to vanish altogether. *Description* of the kind provided by spatial interaction models remains possible longer, until overwhelmed by the structural complexity and the sheer mass of information at levels 5 and 6, not to speak of level 7. The potential for *explanation* within this same framework largely depends on the success we may have in linking this geography of possible worlds with the geography of the world we live in. Other alternative modes of explanation are, of course, not only possible but already flourishing in contemporary geography. Marxian and phenomenological geography appear to be pure-blooded instances of level 6 theoretical ventures, considerably closer to real-world processes than the gravity-model geography of levels 2, 3, and 4, yet in principle just as unreal when it comes to applying their findings in practice. Clearly, abstractions will always be abstractions, and we cannot hope for a theoretical approach that will deal with anything but conceptual constructs: this is the predicament of all science. However, whereas the *natural* scientist can say (Eddington, 1939, page 3):

"Physical knowledge ... has the form of a description of a world. We define the physical universe to be the world so described"

and leave it at that, the *human* scientist cannot just substitute "social" for "physical" and adopt the definition. This circular conception of the physical universe and the ensuing logical closure of the corresponding system of knowledge (though also begging the question of the observer–observed relationship) grants the theoretical constructs within the system a certain autonomy *vis-à-vis* 'external' agents such as human action. Such detachment is clearly extremely problematic in the social sciences, where human theory is constantly interfered with by human action. This is why I think that understanding of the logical status of the theoretical constructs used, that is, of the nature of their relationship with the world of action and experience, is a much more important issue in the human than in the natural sciences. In spite of much brilliant work on the nature and function of models in science, I do not think that this particular question of the logical relationship between what we may call 'the real world' and its various possible theoretical images has been sufficiently elucidated by philosophers. This is perhaps an area where geography, a discipline that spans the spectrum from the purely physical to the intimately human, could make a pioneering contribution to the methodology and philosophy of science (Golledge, 1979).

Third, what are the implications of all this for future research in geography? To my mind, one possible conclusion is that we need more painstaking, unpretentious empirical research on the one hand, and more basic research on fundamental underlying structures on the other. What I think we do *not* need is yet more wishy-washy theories with a semblance of empirical wisdom and a guise of scientific respectability, with which one never knows where assumption stops and concrete fact begins. In particular, we need more *descriptive* work on spatial perception, spatial behaviour, motivation, expectation, belief, habit formation, intuitive processing of spatial information, attitude towards uncertainty, emotive appraisal of the environment, response to environmental stress, and so on, that will explore aspects of level 7 processes. At the same time, we also need more work on *formal* theories that will not only help link these aspects to each other and to the various explanatory frame- works we use, but also clarify the logical relation of the above conceptual constructs to the world of experience (for example, through definition of their domain). It is clear that both these tasks lie well beyond the possibilities of the 'scientific method' and its familiar recipes for data collection, hypothesis formulation, statistical test, and theory construction. They obviously also lie beyond the reach of the 'softer' approaches sometimes used in the social sciences, approaches which all too often smack of scholastic journalism. Peter Gould's plea for the establishment in geographic education of the three 'covers' of mathematics, systems analysis, and philosophy (Gould, 1979a) readily comes to mind.

The need for high-level mathematics, the mathematics of structure even more than the mathematics of quantity, to me at least appears obvious. We may not be able to do without statistics and probability theory, but we shall mainly need to learn more serious geometry, topology, set theory, the mathematics of discrete structures, and possibly some formal logic. And it is not just (or even primarily) a matter of using some new techniques; it is mainly a question of really and thoroughly understanding what we are saying in our various theoretical statements. Most present- day quantitative theories are like fresh eggs—a very thin hard shell on the outside, but all soft and slimy inside. Although we have learned many tricks, the true insight of mathematics, what Heidegger (1967) has called "the mathematical", is still missing from our theories. Since we have become disillusioned with quantitative geography, why not try *qualitative* geography—with a solid base in *qualitative* mathematics?

The need for systems analysis may be more controversial. Since the heydays of systems analysis in the 1960s, criticism against it has reached such levels of perfection that today one feels almost embarrassed even to mention it. However, I think that systems theory still has an important role to play in geography, though not as a theory *of reality*, as has mostly been the case to this day, but as a theory *of theories* of reality, as I attempted to show in these pages. Personally, I do not believe very much in the 'city-is-a-tree' kind of application of the approach. In this area, as in many others in empirical theory-building, many a problem comes from the attempt to map observations directly into the x's and y's of available formalisms without considering the frame of meaning within which the particular theoretical statements may be valid. It is possible that systems concepts could be much more useful when applied to the analysis of our intellectual equipment than directly to the reality we wish to study.

The need for philosophy in geography, which is the theme of this seminar, will be documented, I am sure, very thoroughly by most participants. I just wish to stress the importance of keeping our standards high in an area which could be dangerous because it is so deceptively easy to enter. Indeed, bad philosophy is even easier than bad statistics, and a page is much sooner covered with woolly plagiarism than with an irrelevant model. Quantitative geography at least presupposes a modicum of specialised knowledge; philosophical geography is, in principle, accessible to anyone who can read and write. We cannot expect everyone to be a creative philosopher (*and* a high-powered mathematician, *and* an original systems theorist, *and* a good geographer at the same time), but if we must steer a course in critical geography that stays clear both of the irrelevant and of the trivial, we need a lot more than educated speculation. These are difficult times for geography, and only the best we can do will be good enough.

Geography in philosophy

I want to conclude this essay on philosophy in geography with a few general reflections on the possible philosophical significance of the approach advocated in the preceding pages. Several issues which could stimulate philosophical speculation were hinted at along the way: on the interpretation of notions such as causality, probability, simplification, theoretical assumption, and theoretical law; on the possibility of conciliating opposites such as objective and subjective, abstract and concrete, general and particular, fact and value, analytical and holistic, and so on. Further investigation of these rather technical questions must be left to those better qualified. But there is just one further point of more general concern that may be mentioned here.

Whatever the ultimate fate may be of the particular conceptual framework presented in these pages, the very fact of being able to make meaningful statements about *a priori* structures in an empirical science raises some disturbing questions—disturbing, because they were thought to have been answered long ago. These questions concern the relevance of *rationalism* in (social) science and the role of human rational *understanding* (as opposed to human nature, human action, society, culture, the Institutions ...) in the determination of phenomena observed in the empirical world. We are reminded of Kant's "synthetic *a priori*", the reality due to nothing but the properties of the human mind; we are reminded of Hegel's enigmatic dictum "Reason is the substance from which all things derive their being"; we are reminded of a phrase by Charles Peirce relating the apparent order in the universe to the contingent fact of human intelligence[16]; of Max Weber's concept of "ideal type", the mental construct that is literally true in some logically possible world (Weber, 1949). In more recent years, Wittgenstein based a great philosophy on the premise that it is only through a study of the

[16] Peirce's phrase is "We may, therefore, say that a world of chance is simply our actual world viewed from the standpoint of an animal at the very vanishing-point of intelligence" (quoted in Feibleman, 1969).

logic of factual propositions that the general structure of reality could be apprehended; he therefore set himself the task of the demarcation of the limits of ordinary language (Wittgenstein, 1961; Pears, 1971). The study of the logic of certain groups of *mathematical* propositions, and the demarcation of the limits of certain mathematical languages, is a conceivable extension of Wittgenstein's endeavour.

A significant step in that direction was actually taken by Eddington, who set out to derive some of the fundamental constants and laws of theoretical physics without recourse to observational data, entirely out of the logical properties of the mathematical formalisms and procedures used in their formulation. The degree of success of the venture, which Eddington did not get the time to complete, is still undecided because of the extreme technical difficulty of that work. However, its philosophical implications with respect to the origin of theoretical knowledge are already evident, and they could prove of enormous importance for science as a whole (Eddington, 1939). Moreover, this still isolated attempt must be seen in the context of a more general concern with the possible influence of the process of scientific inquiry upon the result of that inquiry itself[17]. Even in the most 'objective' of sciences, the line between description and construction, between discovery and invention, is becoming very blurred indeed.

This brings us to the final point in this discussion. Throughout these pages there has been talk of models, of definitions, of formalisms, of proofs, of mathematics, of logic; to a number of those who turn to philosophy in the hope of making present-day human geography more human, such emphasis on values and procedures more characteristic of inquiry in the physical sciences may seem a little off the mark. Yet to my mind it is a mistake to insist on maintaining the age-old dichotomy between the natural and the social sciences. It is quite true that further development in the direction advocated here will probably lead to a geography much more akin to theoretical physics than is actually thought likely. This should not be seen as the vindication of old-fashioned social physics, too naive to be of any further consequence, or as a triumph of materialistic and reductionist philosophy. On the contrary, this is a profoundly humanistic perspective which grants the human observer the power to determine the laws of reality, and places the natural sciences at the upper, more superficial levels of a continuous structure of knowledge that reaches very deep. Our search for common ground in the field of

---

[17] It is well known that developments both in relativity theory and quantum mechanics have given rise to such concerns. Heisenberg wrote extensively on this question in connection with his famous 'principle of indeterminacy'; J A Wheeler (1977), in a paper entitled "Genesis and Observership" spoke of the physicist as the "Observer–Participant" —a participant in the shaping of the laws of Nature. Paradoxically, current quantitative approaches in the *human* sciences seem to be beyond such concerns.

geography is motivated by the conviction that there should be no water-tight compartments between research traditions. This applies equally well at the other end of the knower–known relationship: there should be no watertight compartments between the fields of human knowledge. Not necessarily because Everything is One; but because, to deal with everything, *Homo cogitans* has Only One Mind.

References

Angel S, Hyman G M, 1971 "Transformations and geographic theory" WP-72, Centre for Environmental Studies, London; 1972 *Geographical Analysis* 4 350-367

Angel S, Hyman G M, 1976 *Urban Fields* (Pion, London)

Ashby R W, 1956 *An Introduction to Cybernetics* (Chapman and Hall, London)

Atkin R H, 1972 "From cohomology in physics to q-connectivity in social science" *International Journal of Man-Machine Studies* 4 139

Atkin R H, 1974a *Mathematical Structure in Human Affairs* (Crane, Russak, New York)

Atkin R H, 1974b "An approach to structure in architectural and urban design: 1. Introduction and mathematical theory" *Environment and Planning B* 1 51-67

Atkin R H, 1974c "An approach to structure in architectural and urban design: 2. Algebraic representation and local structure" *Environment and Planning B* 1 173-191

Atkin R H, 1975 "An approach to structure in architectural and urban design: 3. Illustrative examples" *Environment and Planning B* 2 21-57

Atkin R H, 1978 "Time as a pattern on a multi-dimensional structure" *Journal of Biological and Social Structures* 1 281-295

Ayer A J, 1972 *Russell* (Fontana/Collins, London)

Batty M, March L, 1976 "The method of residues in urban modelling" *Environment and Planning A* 8 189-214

Blalock H M, 1961 *Causal Inferences in Non-Experimental Research* (University of North Carolina Press, Chapel Hill)

Coleman J S, 1964 *An Introduction to Mathematical Sociology* (Collier-Macmillan, London)

Couclelis H, 1977 *Urban Development Models: Towards a General Theory* Ph D dissertation, University of Cambridge

Echenique M, 1977 "An integrated land use and transport model" *Transactions of the Martin Centre for Architectural and Urban Studies* 2 195-230

Eddington Sir A, 1939 *The Philosophy of Physical Science* (Cambridge University Press, Cambridge); reprinted in 1974 as an Ann Arbor paperback, University of Michigan Press, Ann Arbor

Feibleman J K, 1969 *An Introduction to the Philosophy of Charles S Peirce* (MIT Press, Cambridge, Mass)

Golledge R G, 1979 "Reality, process and the dialectical relation between man and environment" in *Philosophy in Geography* Eds S Gale, G Olsson (D Reidel, Dordrecht) pp 109-120

Gould P, 1979a "Signals in the noise" in *Philsophy in Geography* Eds S Gale, G Olsson (D Reidel, Dordrecht) pp 121-154

Gould P, 1979b "The structure of a discourse space: some pen-on-paper reflections on *Philosophy in Geography*" unpublished paper available from the author, Department of Geography, Pennsylvania State University

Gould P, 1980 "*Q*-analysis, or a language of structure: an introduction for social scientists, geographers and planners" *International Journal of Man-Machine Studies* 13 169-199

Habermas J, 1971 *Knowledge and Human Interests* translator J Shapiro (Beacon Press, Boston)

Harvey D, 1969 *Explanation in Geography* (Edward Arnold, London)

Harvey D, 1973 *Social Justice and the City* (Edward Arnold, London)

Heidegger M, 1967 "Modern science, metaphysics, and mathematics" in *Martin Heidegger: Basic Writings* Ed. D F Krell (Routledge and Kegan Paul, Henley-on-Thames, Oxon) pp 243-282

Hooker C A, 1973 "Metaphysics and modern physics' a prolegomenon to the understanding of quantum theory" in *Contemporary Research in the Foundations and Philosophy of Quantum Theory* Ed. C Hooker (D Reidel, Dordrecht) pp 174-305

Kuhn T S, 1970 *The Structure of Scientific Revolutions* (The University of Chicago Press, Chicago)

Lancaster K J, 1966 "A new approach to consumer theory" *Journal of Political Economy* **74** 132-137

Lowry I S, 1964 *A Model of Metropolis* RM-4035-RC, Rand Corporation, Santa Monica

Lynch K, 1960 *The Image of the City* (MIT Press and Harvard University Press, Cambridge, Mass)

Massey D, 1974 "Towards a critique of industrial location theory" RP-5, Centre for Environmental Studies, London

Nerlich G, 1976 *The Shape of Space* (Cambridge University Press, Cambridge)

Olsson G, 1975 "On words and worlds: comments on the Isard and Smith papers" *Papers of the Regional Science Association* **35** 45-49

Pears D, 1971 *Wittgenstein* (Fontana/Collins, London)

Reichenbach H, 1958 *The Philosophy of Space and Time* (Dover, New York)

Russell B, 1908 "Mathematical logic as based on the theory of types" in *Logic and Knowledge* Ed. R C Marsh (Allen and Unwin, London) pp 57-102

Specht R D, 1964 "The why and how of model building" in *Analysis for Military Decisions* Ed. E S Quable, Rand Corporation R-387-PR, Santa Monica

Weber M, 1949 *The Methodology of the Social Sciences* (The Free Press, New York)

Wheeler J A, 1977 "Genesis and observership" in *Proceedings* Fifth International Congress of Logic, Methodology, and Philosophy of Science Part 2, University of Western Ontario Series in the Philosophy of Science, Eds R Butts, J Hintikka (D Reidel, Boston)

Whitehead A N, 1920 *The Concept of Nature* (Cambridge University Press, Cambridge)

Wilson A G, 1970a "Generalising the Lowry model" WP-56, Centre for Environmental Studies, London

Wilson A G, 1970b *Entropy in Urban and Regional Modelling* (Pion, London)

Wilson A G, 1974 *Urban and Regional Models in Geography and Planning* (John Wiley, Chichester, Sussex)

Wittgenstein L, 1961 *Tractatus Logico-Philosophicus* (Routledge and Kegan Paul, Henley-on-Thames, Oxon)

Questions of historical and social context

The meaning and social origins of discourse on the spatial foundations of society[†]

Allen J Scott

> "Of all cultural worlds, knowledge would seem to be the most detached from social reality. Does it not seem to lay claim to universal validity and to be founded on sound judgements which are usually considered to be the prerogative of individual consciousness?"
>
> G Gurvitch, *The Social Frameworks of Knowledge*, page 3

Introduction
In this essay an attempt is made to decipher the meaning and social origins of theoretical discourse on the spatial foundations of modern (late capitalist) society. The essay seeks to address questions that may at the outset be posed in informal terms as follows: Why do geographers, regional scientists, urban economists, and others study the spatial patterning of social events? Why are some kinds of spatial patterns widely and intensively examined, while others are left in more or less permanent abeyance? Why did the study of the spatial structure of modern society only start to attract really serious academic attention some time after the Second World War? Why is discourse on geographical issues strongly polarized around the competing themes of technique and rationality versus humanism and human values? Why do urban and regional scientists and scholars engage in heated debate about the relative merits of logical positivism and phenomenology? Why do Marxian and neo-Marxian theories of society now begin decisively to penetrate the discourse of urban and regional science? Is all of this intellectual activity an inevitable outcome of the self-propelling logic of scientific and philosophical enquiry? Or is it simply an arbitrary phenomenon describable in terms of some 'sociology of knowledge'?

In this essay, it will be argued in general that knowledge and science (*qua* discourse) are historically-determinate (though definitely not arbitrary) phenomena. In particular, it will be argued that the shape and contents of *urban and regional* discourse are given prediscursively by the problems, needs, interests, and practices that appear in late capitalist society, as prevailing social and property relations are projected through the dimension of geographical space. This is a position that explicitly (and unfashionably) plays down the role of epistemology as a driving

† I want to thank Adil Cubukgil for his many discussions with me on several topics treated in this paper. These discussions were enormously useful in helping me to clarify my own thoughts and arguments. Needless to say, accountability for any specific viewpoint propounded here is due entirely to me.

power in the historical development of modern geographical discourse, and instead affirms that the dynamics of knowledge production in the spatial sciences (as in the case of all other sciences) is governed by determinate social forces. As will be made clear in due course, this position does not necessarily deny certain forms and degrees of autonomy to the levels of theory and epistemology in any given mode of scientific inquiry, though it most certainly does involve commitment to the view that science as a whole is essentially a historically-determinate phenomenon within the overall development of human society.

This project is accomplished in four main phases. First, some basic social and political trends in late capitalist society are examined as a way of identifying the main tensions that evoke specific forms of urban and regional discourse. Second, an attempt is made to elucidate the intricate relationship of such discourse (and its several internal moments) to the overall dynamics and imperatives of contemporary society. Third, a broad effort is made to theorize discourse as a definite social phenomenon. Fourth, and finally, some brief comments are appended on the significance of the resurgence of Marxian and neo-Marxian modes of inquiry in the urban and regional sciences. Note that in this essay, and as a simplifying gesture, there is little attempt to address the dynamics of hermeneutic knowledge within the sphere of spatial and geographical discourse; instead, concern is largely focused on forms of knowing (both technical and humanist) that address current social problems, issues, and realities.

Late capitalism, state intervention, and the emergence of urban and regional science

In conformity with the above agenda, a description is undertaken at this point of current changes in the structure of society, and of the ways in which these changes are expressed in new problems of spatial development. The description provides the main foundations for an investigation of the meaning, content, and structure of urban and regional discourse. Only in the light of an examination of the current historical conjuncture, and of the urban and regional processes (including political and administrative processes) that flow from this conjuncture, is it possible to discuss the specific forces that form and give substance to spatial discourse. We shall then be in a position to interpret current intellectual debates and paradigm shifts in the various spatial disciplines.

The hallmark of contemporary late capitalist society is that production and reproduction relationships are both pervasively managed by the State. The roots of this phenomenon are to be found in the early manifestations of State intervention in nineteenth-century competitive capitalism. However, its decisive and irreversible historical appearance coincides with the first developments of Keynesian economic strategies and welfare-statism some time in the 1930s (Gough, 1979). By the 1930s, and above all in the light of the economic collapse of 1929, it had become abundantly clear that capitalist civil society could no longer secure for itself the basic conditions for its own continued existence. Capitalism could no longer proceed smoothly at high levels of productivity and employment without a very considerable enlargement of the sphere of State control and management. In addition, the State now found itself compelled to supply complex and widely-ranging welfare programmes as a means of under-pinning the whole process of social and cultural reproduction. The State had to regulate the economic cycle, to create conditions for the more effective deployment of underutilized units of capital and labor, to underwrite the costs of commodity production by making available huge quantities of infrastructural services, to stabilize social relationships by means of massive investments in social overhead capital and other instruments of social control, and so on. Furthermore, whereas the new interventionism was predominantly a reflection of nonspatial macrosocial needs and imperatives, it was also in part a response to a buildup of difficult and politically-explosive urban and regional problems, just as it was in part significantly effectuated through urban and regional policies and programmes. Thus, and especially from the time of the Second World War onwards, the State has committed massive and ever-increasing public funds to regional development programmes, highway planning and construction, the building of new towns, urban renewal, public housing programmes, and many other areas of investment. This has had the effect of bringing urban and regional issues more and more into focus, and of conjuring into being a contingent system of analytical, scientific, and policy discourses as the bases of effective political action. It is scarcely cause for surprise, then, that urban and economic geography (as we know them today), regional science, spatial economics, and urban and regional planning began to make their decisive appearance in universities and research institutes just as late capitalist society was starting to get into full swing.

In the decades following the Second World War, Keynesian political strategies and welfare-statism have established themselves as definite and deeply rooted social forms in virtually all of the advanced capitalist societies. Accordingly, a typical developmental pattern of civil and political interaction has tended to occur in those societies. It is a pattern in which critical disturbances in social and economic stability produce various forms of State intervention; such forms of State intervention then unleash new energies which create, in turn, new social problems and predicaments. The new problems and predicaments call for further State intervention—and so on, in a seemingly endless spiral of escalations. For example, in the domain of urban and regional activity, we observe a constant stream of problems in such matters as industrial location, regional integration, urban growth, housing, community development, and all the rest. Whenever these problems threaten the global viability of society as a whole, remedial action on the part of the State is invariably forthcoming. However, by reason of its own functional logic, the capitalist State can only engage in *reactive* forms of planning, and this tends to create the social conditions under which new and more complex sets of problems start to make their appearance (Scott, 1980). Such a process leads irrevocably onward to augmenting State intervention in social and economic life. As Pahl (1977, page 161) has written:

"There comes a point when the continuing and expanding role of the State reaches a level where its power to control investment, knowledge, and the allocation of services and facilities gives it an autonomy which enables it to pass beyond its previous subservient and facilitative role (in relation to private capital)."

In late capitalist society, then, it can surely be said that the basic mechanisms of social regulation are no longer coordinated to any significant degree by a network of market relations, but that, on the contrary, social stability and continuity are nowadays largely secured by means of bureaucratic intervention. In this sense, late capitalist society may perhaps be seen as embodying an incipient form of what we might loosely term a "State mode of production" (Szelenyi, 1981). In this incipient mode of production an ascendant 'new class' of State managers, bureaucrats, planners, professional consultants, academics, and so on, begins to replace the old moneyed class as the main embodiment of social authority and control (Gouldner, 1979). The net result is a society whose central operating principle comes more and more to resemble a managerial process (itself mediated by scientific, technical, and psychosociological discourses).

It must be added at once that the evidently augmenting autonomy of the organs of government does not mean that government becomes simply a selfconstituting historical phenomenon. On the contrary, the more the bureaucratic apparatuses of late capitalist society expand and ramify, the more clearly does their relationship to broader underlying social pressures become evident. In particular (and of major importance for the purposes of the present analysis), the empirical content, style, and objectives of contemporary political decisions are constrained and structured by two very stubborn sorts of social pressures. On the one hand, and in the surpassing interest of economic efficiency and growth, the late capitalist State finds itself having to secure highly rationalized initiatives by means of technical control of resources. At the present time, there is no doubt that this is the dominant mode of State intervention given a prevailing situation in which the fiscal and political penalties of unproductive public expense are indeed severe. On the other hand, the State must also seek to contain the sociocultural stresses and strains that break out at different junctures in late capitalist society, and it achieves this end by means of a proliferation of human relations programmes and social administrative devices. These different and frequently incompatible forms of contemporary economic and social management make their decisive intradiscursive appearance in urban and regional science in the form of a continuing academic-*cum*-policy debate about problems, emphases, techniques, and approaches in the investigation of the spatial foundations of modern society.

The meaning and social origins of urban and regional discourse

Urban and regional science deals with questions that are raised as capitalist social and property relations are projected through the dimension of geographical space. But what, we may ask, determines the specific substantive content of such questions? And what forces shape and reshape the biases, oversights, and theoretical orientations of the answers that are constructed around them? An attack on these puzzles can in principle be launched on the basis of the observation made above, that the geography of late capitalist society is shot through with problems and predicaments which call urgently for escalating remedial control and intervention on the part of government bureaucracies. The connection between an historically-determinate set of problems and predicaments, and a given set of collective interventions, is mediated by definite forms of discourse. Presumably, this means that the forms of discourse bear some discoverable relationship to the concrete social moments between which they come to life. In late capitalism, as we shall see below, this particular relationship is one that leads to a corpus of urban and regional discourse that is deeply split between the perspectives and interests of (a) technical control and (b) human relations management. Let us now deal with these specific perspectives and interests in turn.

First, then, modern capitalism is above all a system of advanced commodity production and accumulation whose workings produce a contingent, but highly problematical, geography. The main empirical substance of this geography involves such matters as regional growth patterns, urbanization phenomena, transport processes, various sorts of land-use configurations, and so on. These substantive matters are of interest to urban and regional scientists to the degree that they are of concern to society generally; in other words, insofar as they underpin the processes of commodity production and accumulation, and, in particular, as they pose *real* problems which threaten (in a variety of ways) the continued viability of society. At the same time, and because these matters are an intrinsic geographical element of the entire structure of commodity production and accumulation, they also participate in the broad rationality of this structure. This implies at once that socially-necessary collective management of the spatial foundations of modern society is above all characterized by forms of technical rationality that ensure (in conformity with prevailing social norms and imperatives), the efficient allocation and spatial deployment of given resources. In their search for efficient and technically rational outcomes, urban and regional planners draw heavily upon technocratic methodologies (such as systems analysis, cost-benefit analysis, large-scale land-use–transport models), and upon positivistic theories of spatial processes (such as neoclassical land-use theory, central place analysis, the gravity model). These methodological and theoretical discourses provide planners with the tools they need for remedial control of the spatial bases of commodity production, and for tackling spatially-mediated breakdowns in the production–wages–consumption complex. Moreover, they are tools that conform precisely to the immanent objectivistic logic of late capitalist society, which is to say a society in which monetary quantities are the pervasive measure of worth and performance.

Second, as capitalist society evolves and becomes more complex, new problems calling for more subtle forms of social, cultural, and psychological management start to present themselves. The new problems emerge once society reaches an historical stage in which the development of human resources becomes a critical and omnipresent public policy issue, and in which complex processes of reproduction and legitimation must be collectively secured. In addition, however, the new problems are in part posited upon the very success of technocratic rationality, which by bluntly resolving issues of physical and economic planning, sparks off further problems which have a more human, personal, and subjective complexion. Accordingly, these new problems now also project themselves into discursive form where they engender ideological effects in the guise of theoretical and philosophical ruminations about such matters as values, subjective meaning, relevance, and so on. For example, the massive and essentially technocratic development of intraurban expressways in North American cities in the 1950s and 1960s destroyed the fabric of many old-established inner-city neighbourhoods. It also caused significant deterioration in the quality of the urban environment, and it was notably insensitive to the disproportionate burden borne by blacks, the poor, the aged, and other disadvantaged groups as their homes were cleared on an unprecedented scale to make way for new expressways. The consequence was a growing sense on the part of many urban denizens of alienation and anger as urban life seemed to become more mechanized and out of control. Out of such feelings as these there congealed a variety of radical urban political movements calling insistently to be heard. The pressures of these movements forced a significant reorientation of much urban planning activity (including bureaucratic internalization of such procedural methods as advocacy planning, citizens' participation, decentralized decisionmaking, and so on), and by the end of the 1960s and the beginning of the 1970s they had succeeded in making municipal administration in North America noticeably more flexible and open in comparison to what it had been in the 1950s and early 1960s. Such pressures and changes were then echoed in various addenda to prevailing versions of the theory of urban planning (Lemon, 1978).

More generally, technical rationality as a modality of social management and control carries society forward to an advanced level of development in which new possibilities of social and cultural existence are unleashed, and in which contingent problems of reproduction and legitimation appear. At the same time, purely technocratic interventions encounter, beyond their immediate terms of reference, affective social relationships and the hostile universe of the subjective and the personal. This confrontation then gives rise to further imperatives of administration and control in order to restabilize the existing fabric of urban and regional society. The State, in late capitalist society, finds itself in the difficult position of having to steer a crisis-fraught path between such frequently irreconcilable and potentially explosive political alternatives. As Habermas (1976) has pointed out, failures to secure technically rational interventions (in matters of resource allocation, productive efficiency, economic growth, and so on) will produce critical breakdowns within the economic system at large. On the other side of the coin, failures at the level of human relations management will compromise the workability of the reproduction system and endanger the legitimacy of the established social order.

Of course, these inescapable 'facts of life' in a late capitalist social formation are markedly present in modern policy and administrative discourses. At the same time, they provide the historically-determinate foundation of an academic and intellectual debate around urban and regional questions. Their intrinsic human interest and political urgency propel such facts of life into immediate consciousness, and in various mediated (and occasionally distorted) ways they reappear within theories of the structure and dynamics of geographical space. Concomitantly, the two-fold imperative of technical rationality and human relations as forms of urban and regional control gives rise to contending theoretical and epistemological positions in mainstream geography, regional science, and spatial economics. Hence, in a society where the this-worldliness and immanence of experience are paramount, there emerges a dominant scientific culture involving technicist policy advocacies, objectivist theories of social reality, and positivism as the discursive counterparts of everyday realities. In relation to the social predicaments and conceptual questions generated by this state of affairs, there now appears a constellation of subsidiary and countervailing 'humanistic' discourses. Far from being absolutely alien to one another, however, the opposing practices, theories, ideologies, and epistemologies, as represented by technical rationality on the one hand and the human relations approach on the other hand, constitute the two-fold face of a single reality. This reality is in essence the imperative in late capitalism of political intervention to secure continued economic progress and social stability. The two faces of this reality express the deeply-engrained problems and managerial tasks that are called into being by the overarching social formation. It is in this sense, moreover, that we can now comprehend the laconic observation of Horkheimer (1976) that intellectual activity is no more than one element in the division of labour, namely a workaday moment in the overall

production of society. Until we comprehend such a situation, a meaningful
and critical transcendence of mainstream urban and regional science is
likely to be permanently postponed.

Practices, knowledge, society
In spatial analysis, regional science, economic geography, and cognate
discourses, knowledge is in essence an effect of ensembles of concrete
social problems and interests. These problems and interests are intrinsic
to the prevailing mode of production, and they accordingly assume a very
definite historical character and form. In late capitalist society, a
proliferation of such concrete social issues exists at the urban and regional
level. As society evolves, so the issues become steadily more complex,
calling for ever more effective and subtle planning interventions.

In saying that knowledge is an effect of social conditions there is no
intention here to suggest that intellectual activity is causally determined in
a simple base–superstructure process resembling one of Thompson's (1978)
mechanistic orreries. On the contrary, intellectual culture is quite free of
any tincture of material causality. Such freedom takes the form of both
(a) the real possibility (and the definite actuality) of the emergence of
perfectly disengaged systems of ideas at any given moment of historical
time, and (b) the massive transmission of intellectual baggage across very
different modes of production and social formations. The point, however,
is that significant human commitment to the nurturing and development
of particular bodies of knowledge tends to be indefinitely adjourned
unless those bodies of knowledge address pressing and specific human
dilemmas calling for investigation and remedial intervention. In line with
this remark, humanly significant but historically-determinate issues
invariably attract immediate intellectual attention. This is as true of the
development of the physical sciences (out of a very basic human interest
in the physical appropriation of nature), as it is of the modern resurgence
of psychological and cultural knowledge as a means of coming to terms
with the puzzles of personal life in a society where masses of individuals
enjoy the leisure to be able to pose this problem to themselves. Similarly,
the discourses of urban and economic geography, spatial analysis, regional
science, and so on start to flourish in scientific terms once the urban and
regional instances have developed to a point of socioeconomic complexity
such that a failure to manage those instances effectively threatens the
very existence of capitalist society as a whole.

Science, in brief, thrives on practical, organizational, and managerial
problems, and these problems in turn emerge out of a given field of
historical action. Hence (and to exemplify the point by means of a
simple case), it is not regional scientists who identify theories of intra-
urban circulation which then inform urban planners that serious transport
problems exist; it is rather the concrete appearance of real transport

problems calling for specific knowledge effects (theories of circulation),
which provide the conditions of existence under which the development
of such theories becomes socially and humanly meaningful. By the same
token, then, urban and regional discourse generally makes its unmistakable
historical appearance when the dissonant effects of geographical space
indeed begin to interpose themselves as barriers to the further progress of
capitalist society. In a similar vein, Mannheim (1952, page 134) has
written: "Nothing can become a problem intellectually if it has not
become a problem of practical life beforehand." The problems of practical
life, then, give rise to intellectual projects in the form of conceptual
problematics, which is to say, integrated systems of ideas seeking to
appropriate in thought, and to explain historically-given domains of
empirical reality. Within established problematics, knowledge advances by
means of methodologically guaranteed cognitive progress in the context of
philosophical legitimations from the theory of knowledge (Habermas,
1971).

   Problematics codify, define, and render coherent specific practical
questions and ideological needs posed from within the concrete structure
of human society. Within their established conceptual boundaries,
problematics ease the tasks of 'normal science' (Kuhn, 1970) by making
possible a socially efficient division of scientific labour in such matters as
hypothesis testing, puzzle solving, empirical description, methodological
clarification, and so on. In such a process, however, problematics take on
a life of their own as they start to posit themselves as autonomic
conceptual logics. As the internal development and refinement of
problematics proceeds, they tend to assume an independent and ideal
existence detached from the specificities of a local historical field of
action. Such detachment, moreover, may become increasingly pronounced
as workers within given problematics confront logical, conceptual, and
philosophical problems that no longer have any definite or necessary
connection to an enveloping system of social relationships. Of course, in
part, such a process is a normal and necessary ingredient in the full
development of any scientific enterprise, and in many cases it leads to
deeply enriched insights as purely logical and theoretical discoveries are
translated back into the sphere of practical and empirical investigation.
But the same process, if unchecked by repeated critical surroundings, can
also lead into the realm of perfectly abstracted metaphysics. Worse, the
elaboration of sophisticated conceptual (ideological) systems that are no
longer fully tuned in to the complexities of social and political life, will
often result in a form of cognitive myopia such that the ideological level
now actively impedes successful scrutiny of existing, and/or changing,
social realities[1].

[1] This form of cognitive myopia is distinguished here from other sorts of ideological
activity whose social role is to universalize and to render legitimate the specific
interests of dominant classes, fractions, and groups.

For example, an elaborate and highly routinized problematic in main-stream economics, with its privileged notions of microeconomic adjustment and equilibrium, obscures and hinders the evident task of reproblematizing contemporary economic realities within the context of pervasive political management (Eichner, 1979). Instead, the mainstream's prior commitment to 'neoclassical' concepts induces it to conjure away the political either with (a) the soporific of welfare economics and the purely utopian policy norm of Pareto optimality, or with (b) the nostalgic and, in practice, unfulfillable call for a return to competitive market principles. However, when the detachment of a problematic from an enveloping social reality proceeds so far that levels of empirical explanation begin to drop significantly, and when practical effectivity is lost, we then observe the beginnings of the classical phenomenon of 'paradigm shift' described by Kuhn (1970), as scientific workers begin to abandon it for work in other more socially productive domains of research. We might say, in fact, that the genesis, development, and demise of paradigms are functions of particular historical conjunctures, rather than of sudden and unpredictable illuminations of the human spirit. In parallel with this notion, Hegel suggests in his great *Phenomenology* that ideas begin life as liberating forces, and evolve in the course of time into suffocating straitjackets.

These various arguments go resolutely in the direction of a conception of discourse as a purely *social* phenomenon or mechanism, like commodity prices, the division of labour, or the family. In other words, discourse is not different and separate from society as a whole, but is indeed fully continuous with the totality of social phenomena. There can be no special dispensation for discourse as somehow lying outside of a specific field of historical action, no matter what (intradiscursive) claims discourse might make for itself as a spontaneously generated entity. Further, just as it is evidently inadmissible to treat one brand of religion as a form of communion with the Absolute while relegating all other religions to the status of anthropological curiosities, so we must definitely treat both 'true' and 'false' discourses as somehow having their roots in given systems of social relationships. Nevertheless, on the basis of all that has been transacted above, we are now in a position to affirm that only discourse that serves and reflects specific historically-determinate human interests (however these may be constituted—for example, technical control, legitimation of the social position of a dominant class, emancipation, and so on) is likely to command the assent and attention of specific groups within society. In the absence of a definite registering within discourse of those interests, we observe the concomitant effect of the reduction of discourse to irrelevance and vacuity. Thus, in a society where the dilemmas of personal life and of human interaction are of more than passing concern, even the naive and incoherent discourse of astrology enjoys a huge success, whereas a substantively correct and conceptually elaborated discourse around the issue of curtains flapping in the wind is likely to be dismissed at once as patently uninteresting. As will be made more clear in the next section, such a reference to astrology is *not* to be interpreted as an advocacy of a purely neutral and relativistic view of knowledge.

As scientific work progresses, and as communities of scholars sort out various conceptual problems at the levels of practices, problematics, and epistemologies, it seems *as if* discourse is born in the domain of epistemology, is then mediated into theories of substantive reality, and then assumes concrete manifestations in the form of definite human practices. This is a purely idealist view of knowledge. Although it is true that there are certainly important and genuine problems internal to and proper to the logical elaboration of theory and epistemology, such a conception of the origins and trajectory of discourse is literally topsy-turvy. On the contrary, only discourses that are posited upon existing problems of social life and practice, and upon existing political interests, stand any likelihood of commanding a significant consensus of scholars and scientists. As a corollary, in such discourses the specific problems that are theorized, and the specific conceptions of the knowledge process that are brought to bear upon the act of theorization, are already given in the substance of prevailing social experience. Above all, insofar as epistemology makes for itself extrahistorical claims to judge the shape and form of knowledge in abstraction from specific human problems, practices, and imperatives, it takes the risk of a headlong flight into dogmatism in the sense that it prematurely judges and forecloses issues that are in essence based in evolving forms of social being. A truly viable prospective epistemology is one that acknowledges, and deals with, its own embeddedness in historical phenomena.

Epistemologies that privilege their own discursive role also set up impenetrable barriers to effective debate. For example, so long as humanist geographers ascribe a preferential discursive role to phenomenological ideas in abstraction from the social conditions that provoke a definite interest in those ideas, they are, by the same token, destroying the intellectual conditions under which they might begin to comprehend (critically, to be sure) the historical meaning and significance of quantitative geographers' investigation of, say, computerized models of transport networks. And so long as quantitative geographers continue to invoke the intellectual authority of an abstracted philosophy of natural science, they are unlikely to see the compelling human interest of, say, Relph's (1976) celebrated phenomenological notions of "place and placelessness". Concomitantly, neither side is likely to discover what it is that *unites* them in their very disunity; namely, the double-sided historical situation in late capitalist society, in which a highly determinate discursive space is created by (a) technical breakdowns in production and growth calling for positivistic knowledge effects and scientifically programmed interventions in order to re-establish the economic order of late capitalism, while (b) associated breakdowns of affective individual and social life give rise to a need for empathetic research programmes and sociocultural management so as to maintain legitimation, smooth reproduction, and cultural continuity. It is the discovery of this treacherous *common ground* that enables us to set about the task of transcending the current debate, and that establishes the main foundations for a new, though as yet largely prospective, system of urban and regional discourse.

In spite of such a promise of transcendence, the contemporary urban and regional sciences flounder in an internally induced intellectual crisis whose root cause is directly related to their conception of themselves as essentially noumenal entities. This conception, however, licenses false consciousness in the form of sterile exercises in convoluted abstracted logics on the one hand, and moralism and subjectivism on the other hand. Then, detached from their essential historical origins, these discourses seek to express their competing ideological viewpoints in the form of ultimate but idealist-utopian policy advocacies. These advocacies are perfectly exemplified in the contrasting positions represented (a) by the theory of rational-comprehensive urban planning, and (b) by the edifying sentimentalism of Friedmann's (1973) *Retracking America*. Such advocacies are idealist-utopian in the specific sense that in the necessary absence of any underlying programme or agenda as to how their normative content may be translated into indicative social and political reality, they must remain no more than unattainable and abstracted *ideas*.

From all of the above, we can retain the following major conclusion: the endemic crisis of economic production and growth in late capitalist society creates the need for specific problematics and policy discourses out of which technical control may be accomplished. But technical control creates an advanced set of social conditions in which a countervailing set of human predicaments makes its appearance—alienation, the destruction of affective human relations, the repoliticization of urban and regional planning, and so on. These predicaments now threaten the stability of social existence. There is accordingly created the need for, and a discursive space within which human relations practices and problematics can proceed. In accordance with Hindess and Hirst (1977) we may now see urban and regional discourse, in all its empirical, practical, and ideological multifariousness, as a determinate form of social activity with its conditions of existence in other forms of social activity. These other forms of social activity (regional growth, urbanization, community development, the destruction of humanly meaningful places, etc), constitute the material foundations of the primary discourses of modern geography, regional science, and urban and regional planning. Around these primary discourses there revolves a constantly changing secondary web of ephemeral, idiosyncratic, and disengaged ideas.

Beyond the current mainstream of urban and regional discourse

In the preceding section, the debate among mainstream theorists as to priorities in urban and regional enquiry and, in particular, as to the competing claims of objective science and techniques versus human values and meanings was shown to be a concrete and dialectically mediated moment of current history. However, it is also a mystified moment in the sense that so long as the participants in the debate fail to comprehend its social roots and historical significance neither convergence nor transcendence of views is possible.

In late capitalism, social relationships are virtually everywhere managed in one way or another by the State apparatus. An immediate corollary of this remark is the observation that the production of geographical space is itself nowadays governed to a significant degree by collective political decisions. The point is powerfully made by Castells (1976, page 80), who commenting upon the urban scene observes that:

"Technical change in industrial societies is leading to a progressive increase in the importance of ... political interventions over other elements of the system. This does not mean that society is becoming more "voluntaristic", but simply that the dominant instance is shifting towards the political as the State progressively becomes not only the centre but the driving force of a social formation whose complexity requires centralized decision-making and control of processes. Consequently, a sociology of the production of space must be increasingly focussed on what is termed urban planning."

With the onward march of late capitalist society, and the expansion of the visible hand of the State as the prime regulator of human affairs, two major issues are raised in the matter of discourse on the spatial foundations of society. In the first place, the specific forms of junction and disjunction between the fragmented discourses of the spatial disciplines and the basic structures of late capitalism are rendered increasingly perceptible. At the same time, it is evident that so long as the mainstream remains theoretically closed off from a comprehension of the social logic of the political apparatus that it unselfconsciously serves, it will never see that its policy advocacies (whether 'conservative' or 'liberal') invariably function as instruments of ideological domination and of political manipulation. In the second place, the expansion and ramification of political apparatuses in late capitalism produce a need for a new discursive synthesis in which the State is fully identified as a concrete social mechanism (rather than simply as an idea or a volition), and whose operating principles are comprehended in their relation to the operating principles of society as a whole. This includes, as Castells suggests in the passage quoted above, a thorough rewriting of urban and regional theory in order to assimilate the dominating effects of planning and State activity.

In point of fact, the emerging State mode of production (with its endemic crisis of stagflation and fiscal demise) has already engendered a variety of quite novel discourses which go in the direction called for at the end of the last paragraph. These burgeoning discourses are far from constituting a homogeneous whole, and they range across a spectrum comprising historical materialist philosophy, neo-Marxism, left Hegelian sociology, radical geography, critical theory, post-Keynesian economics, and so on. By and large, they have in common a strong concern with society as a materialized total structure (as opposed to conceptions of society as an amalgam of individual 'rational behaviors' or 'value systems'); they are all deeply committed to an analysis of the political instance as a historically-determinate and central facet of modern life; they are seriously concerned with questions of class (as opposed to social stratification); and they resolutely run counter to the various idealist interpretations of social reality that have prevailed in mainstream discourses in recent decades. As a corollary, they are immediately capable of theorizing their own historical appearance as socially contingent systems of knowledge.

These new and evolving knowledge effects provide the essential bases of a critical transcendence of mainstream urban and regional analysis. The transcendence is accomplished by means of (a) a demonstration of the organic way in which an internally ruptured mainstream grows out of a single sociopolitical reality which engenders both the empirical content of mainstream discourse and its limited modes of theorization; and (b) a decisive analytical move forward via a more powerful and more widely-ranging (historical materialist, neo-Marxist, left Hegelian, etc) discourse capable of addressing spatial questions in a way that unhesitatingly commands the domains *both* of science and of humanism. The new discourse that is capable of bringing about such a critical transcendence now shows certain though ambiguous signs of its own incipient assimilation into the culture and ideology of the 'new class'—the emerging ascendant fraction of social and political managers committed, as Gouldner (1979) shows, to critical discourse and *ipso facto* receptive to all potentially workable ideas, particularly ideas which promise effectively to theorize the tasks and predicaments of political management itself.

If what has been said earlier about the social dynamics of discourse is correct, such a process of assimilation is scarcely surprising given the overall analytical power, relevance, and social effectivity of the new discourse in a late capitalist society in which bureaucratic intervention and control now constitute inescapable facts of life. A word of caution, however, must be appended to this otherwise optimistic scenario. To the extent that the new discourse continues to insist—in conformity with its own historical antecedents—upon a fundamentally emancipatory vision (hence colliding with dominant social and class interests), to the same extent is it likely to encounter real intellectual and social barriers to its own propagation. But if by contrast it is willing to forego (as much academic Marxist theory now seems willing to forego), this basically radical interest, and address itself by design or by default to problems of overall sociopolitical equilibration within emerging forms of politically administered capitalism, then its further intellectual diffusion seems more than assured. In line with this latter remark, there is evidently good reason for advancing the tentative hypothesis that the new discourse (suitably shorn of its overtly radical trappings) is on the point of sublimating itself into a new mainstream.

Conclusion
Discourse is the medium in which social life unfolds. The unfolding, however, is not unproblematical, for discourse not only reveals but also actively conceals social realities depending on pressures, interests, and imperatives that prevail at any given historical conjuncture. Constant critical vigilance and analytical self-consciousness are essential if discourse is to be made to yield up a full harvest of scientific insights and guidelines for progressive human action. Beyond the current mainstream of urban and regional science, with its ideological blind spots and its never-ending, and in principle endless, internal debates, there can be apprehended a new discourse, capable of problematizing discourse itself and of seizing its intrinsic connections to the course of social and political life. But if we are to be true to a commitment to critical vigilance and analytical self-consciousness, we must now ask what theoretical blind spots does *this* discourse contain, what regressive social purposes may it serve, and in what direction does its transcendence lie? If the analysis in this paper is correct, the definitive answers to these questions are contained within society's future self-discovery of itself.

References
Castells M, 1976 "Theory and ideology in urban sociology" in *Urban Sociology: Critical Essays* Ed. C G Pickvance (Tavistock, London) pp 60–84
Eichner A S, 1979 *A Guide to Post-Keynesian Economics* (Sharpe, New York)
Friedmann J, 1973 *Retracking America: A Theory of Transactive Planning* (Anchor Press/Doubleday, New York)
Gough I, 1979 *The Political Economy of the Welfare State* (Macmillan, London)
Gouldner A W, 1979 *The Future of Intellectuals and the Rise of the New Class* (Seabury, New York)
Gurvitch G, 1971 *The Social Frameworks of Knowledge* (Basil Blackwell, Oxford)
Habermas J, 1971 *Knowledge and Human Interests* (Beacon Press, Boston, Mass)
Habermas J, 1976 "Problems of legitimation in late capitalism" in *Critical Sociology* Ed. P Connerton (Penguin Books, Harmondsworth, Middx) pp 363–387
Hindess B, Hirst P Q, 1977 *Mode of Production and Social Formation* (Macmillan, London)
Horkheimer M, 1976 "Tradition and critical theory" in *Critical Sociology* Ed. P Connerton (Penguin Books, Harmondsworth, Middx) pp 206–224
Kuhn T S, 1970 "Logic of discovery or psychology of research?" in *Criticism and the Growth of Knowledge* Eds I Lakatos, A Musgrave (Cambridge University Press, Cambridge) pp 1–23

Lemon J T, 1978 "The urban community movement: Moving toward public households" in *Humanistic Geography* Eds D Ley, M Samuels (Maaroufa, Chicago) pp 319-337

Mannheim K, 1952 *Essays in the Sociology of Knowledge* (Routledge and Kegan Paul, Henley-on-Thames, Oxon)

Pahl R E, 1977 "Collective consumption and the State in capitalist and state socialist societies" in *Industrial Society: Class, Cleavage and Control* Ed. R Scase (Allen and Unwin, London) pp 153-171

Relph E, 1976 *Place and Placelessness* (Pion, London)

Scott A J, 1980 *The Urban Land Nexus and the State* (Pion, London)

Szelenyi I, 1981 "The relative autonomy of the State or State mode of production" in *Urbanization and Urban Planning in Capitalist Society* Eds M Dear, A J Scott (Methuen, New York) pp 565-591

Thompson E P, 1978 *The Poverty of Theory* (Merlin, London)

Social reproduction and the time-geography of everyday life

"... each individual can only be explained by his position in a particular historical and social structure, which is his ..."

Alastair Davidson

"Human beings have no definite nature until they shape themselves through the concrete activity in which they transform the world"

Richard Lichtman

Amidst the cacophony of clashing conceptual, methodological, and philosophical perspectives characterizing present-day human geography, one new sound is beginning to be heard faintly, yet more and more frequently and with increasing clarity. It is an appeal, to some an unappealing plea, for members of the discipline "to actively integrate human geography with social theory" (Thrift, 1982a). It is a call for "human geographers to reverse their long-time dependence on conceptual impulses from other disciplines, and to move from critique to *active* participation in the theoretical debates and developments now occurring within the social sciences" (Thrift and Pred, 1981). In some of its more precise expressions it is an entreaty for human geographers to address directly that most central and challenging set of questions confronting all of the social sciences and history: the dialectic between society and individual; "the relation between the individual and the collective, one and many, subject and object, I and you, us and them" (Olsson, 1981); the "interplay between individual behavior and experience, the workings of society, and societal change" (Pred, 1981b; cf Gregory, 1978; 1980a; 1980b; Olsson, 1980c; Pred, 1981a).

Among all of Torsten Hägerstrand's considerable achievements, the greatest almost certainly has been his provision of a means by which we, as human geographers and human beings, can think about the world around us and the everyday content of our own lives, and thereby begin to contribute creatively to the modification and elaboration of social theory. For time-geography, as developed by Hägerstrand and his Lund associates, does not merely provide an extremely effective device for describing both behavior and biography in time and space, as well as for conducting accessibility constraint analyses[1]. On the contrary, if one looks beyond the immediately apparent, one finds in Hägerstrand's time-geography a highly flexible language and evolving philosophical perspective

---

[1] The writings in which Hägerstrand has presented and developed time-geography are too voluminous to be fully noted here. However, among the more essential of these items are Hägerstrand, 1970a; 1970b; 1974a; 1974b; 1976; 1978.

whose core concepts of path and project readily lend themselves to dialectical formulations concerning the individual and society[2]. These same concepts, when integrated with other frameworks, make possible a reinterpretation of many of the grand themes of social theory.

An emerging consensus in social theory

As several observers have recently pointed out, there is a new consensus, ultimately, but far from entirely, rooted in the realist writings of Marx, which is emerging among social theorists as well as among historians and philosophers concerned with questions of social theory (for example, Liedman, 1980a; Thrift, 1982a). In essence, adherents to this consensus regard the concrete situations of everyday life as of fundamental importance, because "it is at the scale of actual human practices that a society is reproduced and that its individuals are socialized" (Thrift, 1982a). Since a concern for problems involving ordinary everyday activities, experiences, and consciousness lies at the core of the emerging consensus, and since time-geography is capable of providing rich insights into the details of everyday life, it would appear that one way in which human geographers can enter the lists of social theory is by attempting to integrate time-geography with the various theories of social reproduction and 'structuration' identified with the consensus in question. In order to make an attempt of this nature, as I shall do in this essay, it is necessary to make some preliminary clarifying remarks about social reproduction and structuration, and to summarize critically a few of the more important and recent theoretical contributions connected with those terms.

Society "is not, as the positivists would have it, a mass of separable events and sequences. Nor is it constituted, as a rival school would have it, by the momentary meanings we attach to our physiological states" (Bhaskar, 1979a, page 134). Instead, for any given area over any given time, society may be defined as the agglomeration of all existing institutions, the activities (practices, or modes of behavior) associated with the institutions, the people participating in the activities, and the structural relations occurring between the people as individuals or collectivities, between such people and the institutions, and between the institutions themselves (Radcliffe-Brown, 1940; Berger and Luckmann, 1967; Bhaskar, 1979a; 1979b; Giddens, 1979). If society is viewed in this manner, then the reproduction of society may be defined as that constantly ongoing process whereby, in a given area, the everyday performance of institutional activities (including eating, cleaning, and other mundane practices associated with the institution of the family) results in the perpetuation, in stable or altered form, of the institutions themselves, of the knowledge

[2] The concepts of path and project are briefly spelled out below. Although these concepts can be employed dialectically, Hägerstrand himself has not explicitly done so.

necessary to repeat or create activities[3], of already existing structural relationships, and of the biological reproduction of the area's population[4]. Because social reproduction is inseparable from everyday labor and other practices, it is also inseparable from the reproduction of the material world of buildings, transportation facilities, eating utensils, tools, furniture, and other man-made objects (Bhaskar, 1979a; Thrift, 1980).

Most of the theoretical and analytical literature that is part of the emerging consensus not only portrays the structural relations occurring between individuals, collectivities, and institutions as being prolonged or modified by everyday human actions and material practices, but it also depicts, in one way or another, the same actions and practices as the outcome of structural relations. It is this continually ongoing dialectical interplay between structure and everyday practice that is referred to as, among other terms, 'structuration' (Giddens, 1979). It is worth noting that even though the literature considered here places considerable emphasis on underlying structures, structural relations, and structuration, it represents a rejection, or at least a movement away from, abstract structuralism. This is because abstract structuralists usually confer autonomy upon the existence and operation of their conceptual schemes, thereby mystifying social processes and removing themselves far from any recognition that practice and structure are inextricably intertwined (Strenski, 1974; Giddens, 1976; 1979; Coward and Ellis, 1977; Williams, 1977; Bourdillon, 1978; Gregory, 1978; Kosik, 1978).

The effort to conceptualize "the social construction of reality" made by Berger and Luckmann (1967) may be regarded as one of the earliest expressions of the emerging consensus. Much of Berger's and Luckmann's presentation reduces to a threefold argument. First, *society is the product of human activity*. Second, *society is an objective reality* in the sense that its institutions and associated control mechanisms are experienced by its individual members as undeniable facts that are apart from themselves and impervious to their wishes, rather than as human products. Last, *man is a social product* since, in the everyday performance of the "discrete institutional actions" which cumulatively constitute their biographies, individuals acquire taken-for-granted knowledge about the rules of conduct, the "values and even emotions" appropriate to their particular world, and thereby a body of generally valid truths about reality. That is, "the objectivated social world is retrojected [or internalized] into consciousness in the course of socialization, including the learning of language and the habitual filling of 'institutional roles'" (Berger and

[3] For comments on the relationships between social practice and the production and reproduction of knowledge, as depicted by Marx and subsequent analysts, see Thrift (1979).

[4] Even where biological reproduction is the consequence of sexual activity undertaken outside of the family, its preceding social interactions almost always have their origins in encounters directly or indirectly brought about by institutional activities.

Luckmann, 1967, page 61). Put more succinctly, this threefold line of
reasoning proposes that there is an unending dialectical process by which
society produces men who produce, or create, society. However, what is
missing (among other things) in the Berger–Luckmann formulation is a
spelling out of the detailed means whereby the everyday intersections of
individual biographies with institutional activities, *at specific times and
places*, are rooted in previous intersections, at specific times and places,
yet simultaneously serve as the roots of future intersections between
particular individuals and institutional activities, or the workings of
society.

   Bhaskar (1979a; 1979b), in constructively taking issue with Berger and
Luckmann, has developed a related but somewhat different model of the
individual, society, and everyday practices. In his "transformational model
of society", men *do not* create society "for it always pre-exists them".
Rather, it is an ensemble of ['space–time' variant] structures, [everyday
material] practices and conventions that individuals reproduce or transform,
but would not exist unless they did so" (Bhaskar, 1979a, page 120)[5].
Moreover, although society "does not exist independently of conscious
human activity", society does provide "the stock of skills and competences
appropriate to given social contexts", and hence "the necessary conditions
for intentional human activity (as well as, in any given case, to a greater
or lesser extent circumscribing its form)" (Bhaskar, 1979a, page 120).
Put another way, "men in their social activity must perform a double
function: they must not only make social products but make the
conditions of their making, that is reproduce (or to a greater or lesser
extent transform) the structures governing their substantive activities of
production" (Bhaskar, 1979a, page 122). However, Bhaskar emphasizes
that whereas everyday human activity is intentional, social changes are not
necessarily consciously intended. What Bhaskar readily makes clear is
that in order to deal with the dialectics of individual and society, one
must really deal with material continuity and the dialectics of practice and
structure. All the same, he still leaves us without any conceptual outlet
for depicting how the generation of and consequences of particular
everyday individual practices *at specific temporal and spatial locations* fit
into the continual workings and transformation of society.

   Karel Kosík (1978), the ill-fated Czech philosopher who has lived under
virtual house arrest since 1970, has written on what he terms "the
dialectics of the concrete" and "the metaphysics of everyday life". Kosík
at one and the same time builds upon Lukács's (1976) view that particular
societal phenomena cannot be understood without reference to society's

---

[5] Bhaskar perhaps has not been entirely fair in taking Berger and Luckmann to task
for claiming that man creates society and not recognizing that man always finds
society and its structures ready made. For, among other things, they assert that "the
world ... has a history that antedates the individual's birth and is not accessible to his
biographical recollection" (Berger and Luckmann, 1967, page 20).

"structural totality", and maintains that such a totality is not an empty abstraction, but something which develops and shapes itself out of the concrete everyday practices of people[6]. According to him, it is in and through everyday practices that truly great social changes occur. This is so despite the fact that the consciousness of the human subjects participating in the practices is so marked by the language, products, technology, and other objective remains of reproduction that they are blind to the social activity and relations these things embody, and they dwell under the illusion their day-to-day world is 'natural', an unchangeable reality. In short, individuals are determining as well as determined, the producers as well as the products of history. Despite its clarity, this version of the dialectics of practice and structure also leaves us hanging if we attempt to bridge the gap between general model and determinate situations involving particular individuals at precise times and places. In that sense the question of exactly how the singular sequence of actions and practices making up any particular life history is shaped by society, while at the same time contributing to the reproduction or transformation of that society, remains unanswered.

The "theory of practice, or, more precisely, the theory of the mode of generation of practices", developed by Bourdieu (1977) and employed in works penned with his associates (Bourdieu and Passeron, 1977; 1979), rests on the concept of "habitus". In Bourdieu's ponderous prose (1977, page 72): "The structures constitutive of a particular type of environment ... produce *habitus*, systems of durable, transposable *dispositions*, structured structures predisposed to function as structuring structures, that is, as principles of the generation and structuring of practices and representations which can be objectively 'regulated' and 'regular' without in any way being the product of obedience to rules, objectively adapted to their goals without presupposing a conscious aiming at ends or an express mastery of the operations necessary to maintain them and, being all this, collectively orchestrated without being the product of the orchestrating action of a conductor." Or, somewhat more simply, habitus is "a socially constituted system of cognitive and motivating structures" whose resulting everyday individual and collective practices always tend "to reproduce the objective structures of which they are a product" (Bourdieu, 1977, page 76). Through the operation of habitus the particular economic and cultural practices in which individuals of a given group or class partake appear 'natural', 'sensible', or 'reasonable', even though there is no awareness of the manner in which the practices are either adjusted

---

[6] Kosík, who could not foresee the consensus that was to begin emerging, contended that only philosophical thought can arrive at a truly holistic comprehension of reality. Social theory, for him, cannot capture the "structural dialectical totality" of reality since it is handicapped by the specialist viewpoints of the various social sciences.

to other practices or structurally limited[7]. Thus, in a manner similar to
the formulations of Berger and Luckmann, as well as of Kosík and others,
Bourdieu proposes that everyday practices are synonymous with the
internalization of the external objectified world and the externalization of
internal dispositions, or experience-based thoughts, perceptions, and
'appreciations'.

Bourdieu, unlike the other social theorists previously discussed, makes
explicit note of the role of time and temporal sequencing in practices and
their structural 'determinations', and generally suggests a flowing dialectical
relationship between the practices of particular individuals and the
workings and transformation of society. For example, he asserts: "the
habitus acquired in the family underlies the structuring of school
experiences ... and the habitus transformed by schooling, itself diversified,
in turn underlies the structuring of all subsequent experiences (e.g. the
reception and assimilation of the messages of the culture industry or work
experiences), and so on, from restructuring to restructuring" (Bourdieu,
1977, page 87). Such a scheme is highly suggestive, yet remains lacking
in temporal and spatial detail with respect to the daily intersection of
individual biographies with specific institutional activities.

It is first in Giddens's (1976; 1979) version of the dialectics of practice
and structure, or the process of structuration, that the essential
involvement of "time–space interactions ... in all social existence" (1979,
page 54) is acknowledged, and that the "time–space constitution of social
systems" (1979, page 3) comes to the fore. Giddens introduces the
concept of 'duality of structure', by which he means that "the structural
properties of social systems", whether they perpetuate relative stability or
precipitate change "are both [the] medium and outcome of [the reproduction
of] the practices that constitute those systems" (1979, page 5). Or,
structure is "embroiled in both [the] conditions and consequences [of
action]" (1979, page 49). In Giddens's framework, temporality and time–
space enter into the constitution, reproduction, and transformation of a
social system through a threefold conjunction which expresses and
propagates structure: (1) through the temporal and spatial location, or
'immediate nexus', of specific actions and interactions, or practices;
(2) through the finite life-cycle, or biography, of the individuals undertaking those actions; and (3) through the historical duration and spatial

[7] Bourdieu (1977, page 95) emphasizes that, through the habitus, practices are
governed not by *mechanical determinism* but by consistency requirements, or by the
assignment of limits to invention. "As an acquired system of generative schemes,
objectively adjusted to the particular conditions in which it is constituted, the habitus
engenders all the thoughts, all the perceptions, and all the actions consistent with
those conditions, and no others."

'breadth' of institutions and their attendant activities[8]. Giddens also emphasizes that "via the reproduction of institutions in the duality of structure", face-to-face interaction or social integration "is always the chief prop of" relations between social collectivities, or "the systemness of society as a whole" (Giddens, 1979, page 77). Yet for all his sensitivity to the ever present time–space dimension of society, for all his awareness of the need to link the theory of practice with an "institutional theory of everyday life", and for all his success in interweaving the commonplace exercise of practice and social interaction with the themes of power, practical consciousness (nondiscursive knowledge of the rules of institutions) and ideology, Giddens still fails to provide us with some fundamental answers regarding the details of everyday life. We remain uninformed as to the cement binding the everyday functioning and reproduction of particular institutions in time and space with the actions, knowledge build-up, and biographies of particular individuals.

There is little to be gained by critically summarizing the works on social reproduction and structuration of other authors, such as Touraine (1977) and Williams (1977), who can be identified with the emerging consensus in social theory. Although they occasionally employ rather different foci and categories in depicting the everyday operation of the dialectics of practice and structure, and of individual and society, their conceptual melodies are essentially variations on a theme, and the shortcomings ascribable to them share a now familiar element. In each and every case, the reader is left suspended and unenlightened as to precisely the means by which the everyday shaping and reproduction of self and society, of individual and institution, come to be expressed as *specific* structure-influenced and structure-influencing *practices occurring at determinate locations in time and space*. Such an inadequacy is even characteristic of Wallin's (1980) brilliant effort, inspired by the time-geographic tradition, to formulate a model of "the generative grammar of everyday life"—an effort that disappoints because of its overdependence on biological analogy, its lack of connection to major relevant works within philosophy and social theory (Olsson, 1980a), and its failure to exploit fully the basic concepts of time-geography.

Paths and projects: conceptual keys to the specifics of everyday social reproduction
Through the use of time-geography's core concepts of path and project it is possible, at one and the same time, to dispel the deficiency common to the theories of social reproduction and structuration discussed here, and to contribute to the stream of social theory of which they are a part.

[8] It is significant that at the same time he wrote *Central Problems in Social Theory*, Giddens was aware of some of the ways in which time-geography had developed during the early and middle 1970s.

Time-geography rests on the premise that each of the actions and events occurring consecutively between the birth and death of an individual both has temporal and spatial attributes. Thus the biography of a person is ever on the move with her, and it can be conceptualized and diagrammed at daily or longer scales of observation as an unbroken, continuous path through time–space (Hägerstrand, 1970a; 1970b; 1978; Pred, 1973; 1977; 1978; Lenntorp, 1976; 1978; Thrift, 1977; Mårtensson, 1978; 1979; Parkes and Thrift, 1980; Carlstein, 1981). While unwinding her path in the course of sleeping, contemplating, and participating in everyday practices, the individual is constantly in physical touch with (or in close proximity to) other individuals, other living organisms belonging to the animal and vegetable worlds, and man-made or natural objects, each of which also traces out an uninterrupted path in time–space between the point of its inception or creation and the point of its death or destruction.

In the repeated process of coupling and uncoupling her path with other paths for specific purposes, or in the daily process of forming and breaking up 'activity bundles' devoted to production, consumption, and various forms of social interaction, there are always certain physical, or time-geographic, realities circumscribing the individual's choice of alternatives. The actions, practices, or events that an individual may incorporate within her daily path are limited by her indivisibility, or inability to participate simultaneously in spatially separated activities. They also are constrained by the finite time resources at her disposal each day, by the inescapable fact that all tasks have a duration, and by the fact that all movement between spatially separated points is time consuming (to a degree varying with the transportation technology at her command). The individual's daily path is further constrained by, among other things: (1) the necessity of allocating large chunks of her time to eating, sleeping, and other physiological necessities at fairly regular intervals; (2) her restricted ability to undertake more than one task at a time; and (3) the fact that, because no two physical objects can occupy the same space at the same time, every specific space-occupying facility or piece of land has a limited packing capacity or ability to accommodate events (Hägerstrand, 1975; Pred, 1978; Carlstein, 1981)[9]. Because the path concept stresses the physical indivisibility and finite time resources of the individual, it forces us to recognize that participation alterations in one realm of practice invariably bring participation adjustments or changes in other realms of practice—both for self and for others. It is thus possible to

[9] Legal, economic, social, or class constraints also may impinge upon the sequence of actions, events, and practices which accumulate along an individual's path from day to day. Such constraints are relatable, at least in part, either to the time-geographic project concept, or to time-geography's underlying dialectics; both of these concepts are discussed below.

cast new light on the intimate, intricate interconnectedness of different biographies that is an essential part of the everyday process of social reproduction.

The interconnectedness existing between different particular practices at specific times and places, and between different biographies and the continual workings and transformation of society, is brought into sharper focus through the project concept and its combination with the path concept. In the language of time-geography, a project consists of the entire series of simple or complex tasks necessary to the completion of any intention-inspired or goal-oriented behavior. Whether defined by 'independently' acting individuals, or within an institutional context, the tasks associated with a project almost always possess an internal logic of their own which requires that they be sequenced in a more or less specific order[10]. Like any other project, the preparation of a meal, the organization of a wedding, or the manufacture of a good all require that certain tasks be performed before others. Every one of a project's logically sequenced component tasks is synonymous with the formation of activity bundles, or with the convergence in time and space of the paths being traced out, either by two or more people, or by one or more persons and one or more physically tangible inputs or resources , such as buildings, furniture, implements, and raw materials (Hägerstrand, 1974a; Cederlund, 1977; Olander and Carlstein, 1978; Pred, 1978; 1981a; 1981b; ERU, 1980; Carlstein, 1981).

Inasmuch as each of the institutions composing society is equatable to its everyday and longer-term practices, it may be alternatively stated that each of society's component institutions is synonymous with the everyday and longer-term projects for which it is responsible. If this is so, then *the details of social reproduction, individual socialization, and structuration are constantly spelled out by the intersection of particular individual paths with particular institutional projects occurring at specific temporal and spatial locations.* Such intersections may be fleeting, as when an individual's daily path is channelled through the physical facilities of a theatre or a store, or some other cultural or economic institution in order to carry out a consumption project. But such intersections may also be of longer duration, and spread out over a number of times and places, especially when an individual's life path becomes linked up with a specialized and independently existing role within an institution; for example, a student within a university, an employee within a factory, a

---

[10] For some important differences between those projects defined by independently acting individuals, and those defined within an institutional context, see Pred (1981a; 1981b). Also note the very different project conceptualization contained in Schutz's (1967) constitutive phenomenology.

pastor within a church, an administrator within a public agency, a soldier within an army, and so on[11].

If this view of everyday matters is to provide any meaningful contribution to the theory of social reproduction and structuration, it remains necessary to demonstrate, in time-geographic terms, how the details of everyday life—the intersections of individual path and institutional project, as well as the individual execution of extrainstitutional projects—at one and the same time are rooted in everyday details of the past, and how these serve as the roots of everyday details in the future. It is necessary, in other words, to demonstrate how the dialectics of detailed situations that underlie time-geography are relatable to structuration, or to the dialectics of practice and structure.

Two time-geography based dialectics[12]

Within the context of the *external* (corporeal action)–*internal* (mental activity and intention) *dialectic*, each intersection of individual path and institutional project, each temporally and spatially specific individual contribution to social reproduction, is not considered as an isolated event, is not viewed as a disjointed confrontation between the subject and her surrounding world. Each such intersection is part of an integrated and continuous flow of the external or corporeal, and the internal or mental. For as a person incessantly pushes ahead in time–space along the tip of an always advancing 'now line', where becoming is transformed into passing away, she is at the center of a repeated dialectical interplay between her corporeal actions and her mental activities and intentions, between what she physically does and what she is able to know and think[13]. On the one hand, *as an individual traces out her physically observable daily and life paths, corporeally participating in institutional* (and independently defined) *projects, and thereby interacting with other persons and objects, she inevitably amasses internal impressions and experiences that are fundamental to her absorption of normative prescriptions and rules and to the shaping of her beliefs, values, perceptions,*

---

[11] The specialized roles embedded within institutions are usually independently existing—until terminated or superceded—in the sense that when they are not filled by one person they sooner or later must be filled by another. However, note that the project obligations and prerogatives connected with any given, independently existing, institutional role need not necessarily be carried out or exploited in exactly the same manner by the succession of individuals who bring their life paths to that role.

[12] Owing to length limitations, and the overall orientation of this essay, some other dialectics underlying time-geography are not considered here. For a discussion of the path-convergence–path-divergence, or creation–destruction, dialectic see Pred (1981a).

[13] See, for example, Thrift's observations on ideology and 'unknowing' (Thrift, 1979; Thrift and Pred, 1981).

attitudes, competencies, expectations, tastes, and distastes [14], *and her conscious and subconscious motivations, and hence her conscious or subconscious goals and intentions* (only some of which will be realized). On the other hand, *in the process of adding actions to her daily and life paths, or in the process of choosing among and carrying out institutional projects, and of defining and executing personal projects* (that is, in the process of choosing among, and participating in, activity bundles that are possible within the environment's time-geographic constraints), *an individual cannot escape the influence of her previous mental impressions and experiences and consequently derived goals and intentions as well as practical knowledge* [15]. Or, in Olsson's non-time-geographic language (1980b, page 23): "thought-and-action is not something merely physical and observable. Neither, of course, is it something merely mental and nonobservable. It is instead ... one turning into the other and the other turning into the one."

The external–internal dialectic can be rephrased largely in the following manner. When an individual's daily (or life) path is steered through specific temporal and spatial locations as a result of involvement in, or intersection with, a particular institutional project, she is confronted by environmental impulses, personal contacts, influences, and information in general, as well as emotions and feelings, that she otherwise would not have experienced internally. Moreover, her practical knowledge, her awareness of the 'reasonable' and the 'unreasonable', her unarticulated sense of limits, are embellished or reinforced in a manner that otherwise would not have occurred. When an individual consciously or unconsciously employs the mental experiences and practical knowledge acquired through participation in institutional projects to formulate her goals and intentions, and to choose among alternative possible institutional (or independently defined) projects, she sooner or later engages in corporeal action that otherwise would not have become a part of her daily and life paths. In this connection, it is important to realize that even when an individual physically undertakes an independently defined project, or a personal project unassociated with the simultaneous workings of the family or any other institution, she can rarely, if ever, completely evade the mental imprint left by her past participation in institutional projects. For, if an

[14] In this connection Bourdieu (1977, page 124) speaks of the "socially informed body", with its socially determined senses: "the sense of necessity and the sense of duty ... the sense of balance and the sense of beauty, common sense and the sense of sacred, tactical sense and the sense of responsibility, business sense and the sense of propriety, the sense of humour and the sense of absurdity, moral sense and the sense of practicality, and so on".

[15] See, for example, Giddens's (1979) concept of the 'reflexive monitoring of conduct'. He emphasizes (1979, pages 41-42) that intentionality is a continuous flow, that "intentions are only constituted within the reflexive monitoring of action". See also Parkes and Thrift (1979) on the role of predispositions and preconditions in the emergence of individual paths, and footnote (20) below.

independently defined project, such as reading a book or polishing shoes, is motivated by nonphysiological 'wants' or 'needs', these mental dispositions must be recognized as culturally arbitrary, and thereby only acquirable through socialization or previous encounters with institutional projects. Furthermore, if an independently defined project, such as eating, passing bodily wastes, or sleeping, involves the satisfaction of physiological needs, then the project's actual details are very apt to be conditioned and circumscribed by culturally arbitrary notions regarding proper body position, appropriate general time and place, and suitable objects to be employed—notions mentally absorbable only via past involvement in institutional projects. For example, whether one normally eats sitting on a chair or cross-legged on the floor, or passes waste in a seated or squatting position, is a cultural arbitrary. "Suitable objects to be employed" would, of course, include specific foods to be used when eating.

The operation of the external–internal dialectic does not only mean that every determinate intersection of individual path and institutional project is both rooted in the past and potentially rooted in the future. Since the physical enactment of each everyday practice cannot be separated from the influence and modification or reinforcement of mental activity, it also means that the corporeal unfolding of an individual's path in a succession of projects is synonymous with the constitution, development, and expression of consciousness (Ley, 1978). Since, by definition, the accumulated project-participation elements of two individual paths cannot be identical in their time–space details (note the first and last of the time-geographic realities mentioned on page 164 above), each person's physically observable life path, or biographical contact with objective reality, is unique, and each person's development and manifestation of consciousness must also be unique (Weintraub, 1978). However, insofar as individuals who live in the same area have a common class or socio-economic background and belong to the same generation, they are likely to have amassed numerous similar or common path elements. Thus, the uniquely emerging consciousness of many of them may contain strong ideological resemblances, as well as a shared 'structure of feeling' or sense of belonging to the particular social formation in which they find themselves (Williams, 1961; 1973; 1977; 1979)[16]. In other words, through participation in *some* of the same types of projects, through media exposure to and discussion of *some* of the same political–historical occurrences, celebrities, and popular-culture events, and through project-based contact with, or exposure to, manufactured objects of the same

---

[16] Because he is "concerned with meanings and values as they are actively lived and felt", Williams (1977, page 132) distinguishes "structures of feeling" from the "more formal concepts of 'world-view' or 'ideology'".

brand, the symbolic systems belonging to the consciousness of such individuals may be capable of evoking many similar associations[17]. Through the external–internal dialectic, *the ordinary individual is not only created* by *society*, or socialized, *but creates herself*, purposively or habitually adding action elements to her path by internally reflecting upon, or in other ways drawing upon, what she has been externally exposed to, *thereby contributing* (usually unknowingly) *to social reproduction and the perpetuation or transformation of society's structural relationships*[18].

The role of the external–internal dialectic in social reproduction is even more far-reaching than is immediately apparent. The particular institutional projects which are everyday made a part of particular individual paths are not the product of autonomous forces. They do not spring up fully grown out of nothingness. They are not *givens*. Instead, their component activity bundles and roles are a result of the goals established and decisions reached either by the separate individuals, or coalitions of individuals, who hold power and authority within institutions. That is, the definition of every institutional project is in some sense a consequence of the external–internal dialectics associated with one or more particular business executives, administrators, organizational leaders, public officials, parents, or other holders of institutional power and authority. The internal inputs and inventiveness used in the formulation of project-determining goals and decisions are always unmechanistically derived from the previous external path and project-participation record of the goal-setters and decisionmakers themselves. Equally, the project-defining goals and decisions arrived at by individuals holding institutional positions of varying importance cannot be divorced from (1) the 'environmental images' such people have built up through their limited acquisition of imperfect information (Persson, 1977; Törnqvist, 1979; ERU, 1980), (2) the way in which they interpret and react to political and economic events occurring outside the institution, (3) any anticipation they may hold of awards or penalites, or (4) from their absorption or rejection of prevailing values and norms, all of which, in turn, cannot be divorced from their own uniquely accumulated path history, or intersections with the workings of society[19]. Especially both within the family and today's larger-scale economic and governmental organizations, the segments of a power-wielder's earlier path falling within

[17] Such exposures or discussions are in themselves always identifiable with projects of one type or another.

[18] Performance of habitual or familiar acts seldom requires reflection. "Memory traces" of "how things are to be done" (Giddens, 1979, page 64) are more likely to be called upon.

[19] Because of its dialectical context and its emphasis on antecedent societal influences, or on the continuous flow of behavior in time and space, this formulation of the role of individuals in shaping institutions is in no way to be confused with the 'voluntarism' of either Max Weber, Talcott Parsons (1949), and Karl Popper (1966), or any of their followers.

the institution in which he or she holds sway are apt to exercise a complex
and strong (but normally hidden-from-self) effect on his or her consciousness
and way of thinking, and thereby on the project-defining goals that person
reaches singly or in collaboration[20].

The means by which the details of everyday life can be rooted in past
intersections of individual path and institutional project, and also
simultaneously serve as the potential root of future intersections, are
expressible in terms of another dialectical relationship—the *life-path*
(biographical)-*daily-path* (everyday) *dialectic*. This dialectic is closely
related to the external–internal dialectic, but it deals somewhat more
explicitly with the interplay between the long-term commitment and
everyday practice, between a person's lengthier institutional role connections
and her daily-life content.

The general contours of a person's biography are usually depicted in
terms of the overlapping succession of long-term institutional roles with
which she becomes associated—both in a family context (daughter, sister,
wife, mother) as well as in others (for example, as a student at a specific
educational institution, as a practitioner of a particular job, trade, or
profession, or as a member of a given noneconomic organization). Such
lengthy associations are interwoven with everyday life in a reflexive and
subtle manner that most often goes unnoticed by the person herself (as
well as the bystanding observer or biographer). When an individual's
biography or life path becomes linked either with a family role, or with
an independently existing specialized role within some other institution
[see footnote (11)], she must intermittently channel her daily path to
activity bundles belonging to certain routine or nonroutine projects. Yet
when she participates in such activity bundles at precise geographic, and
more or less fixed temporal, locations, *she has her participation in other
types of activity bundles and projects constrained*. Thus, each day she
undergoes experiences, interacts with other people, encounters symbolically
laden inanimate objects, as well as first-hand and second-hand ideas, and
more general information impulses that otherwise would not have come
her way in exactly the same form. These sorts of experiences, interactions,
and encounters help her to define and redefine herself, to renew and
initiate strengths and weaknesses (Erikson, 1975), and to crystallize the
choices she subsequently makes about other long-term institutional roles
she may wish to seek. This is not simply because it is in wandering over
daily paths that people intentionally or unintentionally learn of the life-
path opportunities society has to offer. As the external–internal dialectic
suggests, it is also because it is out of the accumulated sediment of
everyday project-based interactions and reactions that role competencies,
long-range intentions, and choice motivations emerge (cf. Cullen, 1978).

[20] This view is consistent with interpretations, at least partly influenced by Freud, of
family dynamics (Poster, 1978).

Hence, the life-path–daily-path dialectic, which implies essentially that *no institution can touch an individual's life path without influencing the course of her daily path, and no institution can influence the course of her daily path without having the potential to touch upon her life path.*

The operation of the life-path–daily-path dialectic means, in essence, that there is a constant dialectical relationship between, on one hand, an individual's previous assemblage of institutional roles and her resultant record of everyday project participation, and, on the other hand, the objective life-path opportunities that remain open to her (Wallin, 1974). At any point in adulthood the long-term institutional roles that a person may realistically choose among are, in whole or in part, hemmed in and thrown open by the manner in which her life-path–daily-path dialectics have evolved since birth[21]. At birth the infant possesses "built-in primitives" which enable her to suck, grasp, and babble, but her mind is a *tabula rasa*, ready for anything[22]. But as her childhood path is set in motion, and as she assumes the role of daughter and becomes wrapped up in the projects of a particular family at a particular time and place, her mentally virgin condition immediately begins to be assaulted. She begins to be socialized in a given way[23]. She also acquires a characteristic body language (Bourdieu, 1977), and a certain practical mastery of self-care. She learns a given spoken language or dialect, as well as its attendant ways of classifying and understanding information and experiences (Bernstein, 1975a; Hawkes, 1977). She becomes familiar with certain cultural or subcultural conventions and traditions. She develops a sexual identity, and notions regarding the sexual division of labor (Erikson, 1963) that are much influenced by the constraints that production and consumption projects place on the family participation of her parents

[21] However, the degree of selectivity associated with a given institutional role category also may prove constraining. For example, macrolevel economic conditions may result in otherwise realistic choices being closed off, by causing a reduction in the availability of roles of a particular type, and an imbalance between the number of open roles and the number of prepared or qualified role-seekers. See, for example, Hägerstrand (1972); Ellegård et al (1977); and Pred (1973; 1978) on the time-geographic 'matching process' between 'population systems' and 'activity systems'.

[22] In more technical language, "at birth the general recognition and association capabilities of a [person's] neural network function with a minimum of bias and are as content to make one association as any other" (Cooper, 1980, page 15).

[23] Since socialization, by definition, occurs through involvement in the workings of society, or participation in institutional projects inside and outside the family, it coincides with social reproduction and transformation and thereby continues throughout one's life, rather than ceasing at maturity. Moreover, to the extent that parents or others are mutually involved in projects with a child, the socialization resulting is not just that of the child—conduct influences also move from child to parent or other project participants, and not merely in the opposite direction. See, for example, Giddens (1979, pages 129–130); Berger and Luckmann (1967, page 137); and essays contained in Richards (1974).

(Pred, 1981a; 1981b). In addition, her self-image, emotional disposition, and personality emerge uniquely owing to her singular succession of daily-path encounters with her parents, siblings, relatives, and friends[24]. Such everyday, early childhood encounters are also synonymous with the acquisition of elementary learning skills, as well as the instillment of a selected range of values, attitudes, norms, preferences, and taboos that are culturally and subculturally arbitrary, but that appear fully necessary and natural because of their prior existence. In the very long run, or once an individual's 'dependent' stage of life has come to an end, a tremendous number of daily-level project-based incidents, having contributed to the formation of such personal attributes, may consciously or subconsciously enter into the development of wants and intentions that influence the successful or unsuccessful seeking of specific long-term institutional roles.

More immediately, in modern societies, the receptivity to knowledge, the capacity to assimilate, and the performance drive that a young person brings to her first long-term role as a student at a particular school depends on the detailed daily-path content of her earlier and continuing family role (Bourdieu and Passeron, 1977). In turn, her participation in the daily projects required of primary school attenders, along with a particular mix of other students and teachers, marks a new phase of value and norm internalization, and it either succeeds or fails in preparing her emotionally, qualifying her intellectually, and predisposing her for the long-term role of student at a specific secondary school[25]. At secondary school, too, personal history, socialization, and social reproduction proceed in parallel with the individual's daily path. There too, that daily project involvement, made unique by the coparticipation of a given set of teachers and fellow students, either does or does not prepare, qualify, and predispose her for the long-term role of student at a specific university or professional training establishment. And so on, until the individual's string of long-term educational roles is terminated and, on the basis of daily acquired information and experience, she freely or grudgingly opts for one or more occupational or other long-term institutional roles for which she is qualified, or to which she is perhaps attracted (see Bourdieu quoted on page 162 above; Hoppe, 1978).

While the life-path–daily-path dialectic involves an interplay between the individual's accretion of competencies and dispositions and her entrance into new long-term institutional roles, the more fine-grained information and interaction aspects of that dialectic should not be lost

[24] See Mead (1934), and Berger and Luckmann (1967), regarding the genesis of the self and the encounters that occur from birth with the "significant others who are in charge" of a child's socialization.

[25] Note, also, the simple fact that the particular primary school a child attends, and therefore the particular experience she obtains in her first student role, depends on the residential locus of her family role and projects.

from sight. When an individual voluntarily undertakes to move from one employer to another, it is almost always ultimately owing to job-opening information encountered verbally or in print while fulfilling the project demands either of her current job or some other already held institutional role. When a person gets married, and thereby commences a new and presumably long-standing family role, it is inevitably finally due (except in the case of a marriage contract) to an initial daily-path interaction—an interaction most likely to have been directly or indirectly brought about by the activity bundle requirements of an institutional project. When an individual decides to join a given philanthropic, political, or professional organization it usually is in some measure ultimately because of a temporally and spatially specific personal contact that arose in conjunction with the carrying out of some role-dictated institutional project task.

The life-path–daily-path dialectic takes on much added significance if it is seen as simultaneously operating for numerous individuals in the same area or country. It then also helps to illuminate the dialectical links between the everyday details both of institutional functioning and individual paths and of the reproduction of class and *institutionally based* class relationships (Peet, 1975; Giddens, 1973; 1979, pages 109–110).

The early socialization that results from filling the role of daughter or son daily in a particular family is not haphazard in its content. Instead, it is highly dependent on the socioeconomic background of the child's parents. In other words, that content is itself primarily a question of the way in which the life-path–daily-path dialectic has channelled the child's parents through a succession of long-term institutional roles, and the way it has imbued them with the particular linguistic attributes, values, norms, and perspectives that they can pass on to their offspring through everyday family projects. Thus, as a result of their own childhood and subsequent extrafamily institutional socialization, parents of similar background tend to produce children possessing similar dispositions and perspectives on the taken-for-granted world about them [26]. Intergroup, or interclass, dispositional dissimilarities are usually reinforced or magnified on the whole (but not for every single individual) via the project participation necessitated by long-term student roles; not least of all because of the varying 'distances' separating the kinds of dispositions, subcultural baggage, and linguistic skills already borne by children of different groups, and the kinds of dispositions, culture, and linguistic skills educational institutions expect and attempt to inculcate (Bernstein, 1975a; 1975b;

[26] Because of unique life-path interaction characteristics, children with essentially the same "lower-class perspective" on the world may develop "a mood of contentment, resignation, bitter resentment, or seething rebelliousness" (Berger and Luckmann, 1967, page 131). Note that the process of childhood cognitive development and early socialization suggested by the life-path–daily-path dialectic differs from the now classic theories of Mead (1934) and Piaget (1955) because, among other things, it explicitly acknowledges an institutionally (and thereby class) differentiated society.

Bourdieu and Passeron, 1977; 1979)[27]. The day-to-day level of
scholastic achievement normally influenced by these distances either
creates or reinforces already formed subjective expectations as to future
educational and career possibilities. Finally, because of this, and because
subsequent admission to long-term roles within higher educational and
employment-providing institutions is determined largely by performance in
a sequence of previous student roles, children of a given background tend to
be shut out of, or welcomed into, the same general types of privileged,
economically remunerative, and power providing roles as their parents[28].
In short, owing to the cumulative consequences of the life-path–daily-
path dialectic, which are temporally and spatially singular and therefore
unmechanical, an individual is never merely a child of her own time.
Even while actively and intentionally cutting her own swath, she is also in
some measure a child of her own place and social formation.

Dominant projects and the dialectics of practice and structure

We have seen that there is an emerging consensus within social theory
which argues in various ways that social reproduction, and the individual–
society relationships it subsumes, occurs via the process of structuration,
or the dialectical interplay between structure and everyday practice. I have
also contended that the details of social reproduction, socialization, and
structuration constantly take the form of temporally and spatially specific
intersections between individual paths and institutional projects, and that
these intersections are inseparable from the repeated personal working
out both of an external–internal dialectic and a life-path–daily-path
dialectic. Put another way, if social reproduction is always temporally
and spatially specific and tied to individual-level dialectics, then, by
definition, the dialectical bonds connecting everyday practice and the
structural relations between individuals, collectivities, and institutions are
also always temporally and spatially specific and conjoined with individual-
level dialectics. Such a formulation, however, demands further elaboration,
since it is clear that not all institutional projects are equally important to
the time–space specific interplay between structural relations and everyday practice.

In any society the most significant structural relations existing among
individuals, collectivities, and institutions are ultimately traceable to the
activities or practices engendered by those institutions which occupy
positions of domination within that particular society. Domination, which

[27] The projects of educational institutions also contribute to the prolongation of
intergroup dissimilarities to the extent that their content succeeds in legitimating the
power-relationships characteristic of a society. See also footnote (28).
[28] Particular individuals of a given class background may, of course, achieve much
more or less in educational or career terms than might be expected, because of singular
abilities, exceptional contacts outside the home, or unusual family circumstances.

frequently is associated with the exercise of power and the control of resources, acquires an additional dimension if it is recognized that the control of resources extends to the control of *time*, as well as of the more familiar space-occupying material resources (Giddens, 1979). Since all individuals are indivisible, and since individuals singly and collectively have finite daily time resources at their disposal, it follows that institutions deliberately or undeliberately compete for the usage of exclusive times at particular places (Hägerstrand, 1977; Parkes and Thrift, 1979). As a result, those institutions occupying positions of societal domination are those whose projects are dominant, either in the sense that they take time-allocation and scheduling precedence both over the projects of other institutions and over extrainstitutional individually defined projects[29], or in the sense that the time resources they demand force some other projects to be pushed aside totally and obliterated—along with any traditional skills and knowledge necessary to their performance. In fact, the structural relations attendant upon institutional domination can neither exist, nor be reproduced or transformed, in the absence of dominant projects, or activity routines. In the early industrial capitalist societies of Europe and North America, the accumulation-motivated projects of large-scale workshops and locally owned factories were dominant in urban places (Thompson, 1967; Pred, 1981b; Thrift, 1982b). In present-day advanced capitalist societies, the dominant projects of most locales are those of large-scale, corporately owned, and distantly controlled business organizations, and those of economy-intervening and service-providing organizations that are part of the state apparatus. In other societies, both past and present, other economic or noneconomic (for example, religious) projects have been dominant.

The explicit or implicit rules underpinning dominant projects require that participating individuals expend their labor power, or in some other way engage themselves, in activity in a given manner at a given time and place (or institutional domain), rather than doing something else somewhere else during the same time period[30]. Although they may leave some room for individual autonomy or options in the execution of component tasks (see, for example, Burawoy, 1979), dominant projects therefore place certain individuals and groups of individuals in a subservient, dependent, and conflict-laden relationship with other power-wielding individuals and groups, who define the projects in question, and who either own, or have jurisdiction over, whatever material or wage-paying resources are necessary

---

[29] Dominant projects also may take time-allocation and scheduling precedence over secondary projects within the same institution.

[30] The rules underpinning dominent projects *and all other institutional projects* themselves only can come into being, and be maintained or reproduced, via the intersection of individual paths and institutional projects. Their origins cannot be separated from the historically specific institutional contexts within which the past life paths of the rule-makers have unfolded.

to project completion. Dominant projects further affect structural
relations by greatly contributing to the 'local connectedness' of events,
biographies, and collateral processes, or by setting time and place
constraints on project participation and social interaction within the
family and other nondominant institutions (Pred, 1981b)[31]. They also
play a prominent part in the external–internal and life-path–daily-path
dialectics of those who partake in them[32]. Insofar as dominant projects
carry over effects into other, often seemingly unrelated, realms of social
activity and individual experience and consciousness, it is important to
emphasize that material-production or economic projects have not
necessarily been dominant at all times and places. For "although it can
be established *a priori* that material production is a necessary condition
for social life, it cannot be established *a priori* that ... it is ultimately
determining of the rest of social life" (Bhaskar, 1979a, page 126).

   In some instances dominant projects both contribute to and express
intergroup structural relationships through the establishment of 'authority
constraints', or legally, economically, or otherwise based restrictions
regarding who may be permitted to take part in such projects at appointed
times and places. Temporally and spatially specific dominant projects
also enter into temporally and spatially specific structural relations,
inasmuch as they are synonymous with the existence of opposing and
incompatible interests, or with tensions and conflicts between institutions
(and affiliated groups) that succeed in generating such projects and those
that fail at doing so. Furthermore, when in the course of social
reproduction numerous similarly defined and motivated dominant projects
are simultaneously implemented by institutional units of the same type
(for example, firms), they may precipitate social contradictions, or
unintended results which have an impact on the structural relations between
the institutional units themselves, as well as between the institutional
units and affected individuals and collectivities (Elster, 1978; Giddens,
1979)[33].

   Since social reproduction and the dialectics of practice and structure
can only be expressed through an unending sequence of temporally and
spatially specific intersections between individual paths and institutional

[31] Or, in the terminology of Parkes and Thrift (1979), dominant projects 'entrain'
that portion of project participation and social interaction occurring in connection with
nondominant institutions. Note, also, that when dominant projects involve the
payment of wages, the magnitude of those wages affects structural relations by placing
paying-ability constraints on employee participation in the fee-demanding projects of
various types of nondominant institutions.
[32] Given the structural consequences of dominant projects expressed in this and the
previous sentence, one can well appreciate Giddens's view (1979, page 115) "that class
above all refers to a mode of institutional organization".
[33] Among other examples, Elster cites the further decline in profit rates which results
when numerous firms attempt to stem already falling profit rates through the
(dominant-project) use of labor-saving devices.

projects, social transformation and altered structural relations can only occur through the introduction, disappearance, or modification of institutional projects. In other words, social change and altered structural relations can only appear through the addition, elimination, or recasting of the temporally and spatially specific path couplings demanded by institutions from day to day. As a corollary, the most far-reaching social transformations and changes in structural relations come about through the inauguration, discarding, or significant adjustment of dominant projects. This is true whether dominant projects are initiated or otherwise affected owing to (1) the unintended contradictions and crises resulting from earlier project implementation; (2) the intergroup conflicts intrinsic to previous project enactment; (3) capital accumulation and other power-yielding motivations; (4) imposition or radical disruption from outside a society; (5) natural catastrophe; (6) information exposure and the diffusion of technological or other innovations; or (7) to some more or less complex combination of such circumstances. In any event, whether or not the inception, abandonment, or modification of a dominant project has simple, complex, or overdetermined origins, the inventiveness, goals, and rules brought to bear by institutional decisionmakers are indissolubly connected with their own external–internal and life-path–daily-path dialectics, and hence with the dialectics of practice and structure of which they are a part[34]. As I have suggested earlier, this also holds for the inventiveness, goals, and rules brought to bear on the introduction, elimination, or redefinition of all nondominant institutional projects which, on a day-to-day basis incrementally, without drama, and unbeknown to their perpetrators, contribute to the making of history, to social transformation, and the mutation of structural relations[35].

Inasmuch as the everyday carrying out of any institutional project, and especially any dominant project, automatically constrains the possibilities of its participants taking part in other projects, such 'coupling' constraints must always appear with the introduction of new institutional projects, and disappear with the elimination of previous institutional projects. Moreover, the redefinition of existing institutional projects is apt to result either in a tightening up or in a relaxation of coupling constraints. Hence, coupling constraints *and possibilities* are never merely a given. If they are the product of the initiation, termination, or modification of institutional projects, they are also, by definition, the product of the dialectics of

---

[34] This observation has a particular meaning when the appearance or disappearance of a dominant project (or any other institutional project) is at least partly attributable to innovation diffusion. For, the information circulation and exchange generally acknowledged to be so central to innovation diffusion processes is largely dependent on the project-based coupling of individual paths in time and space.

[35] See, for example, pages 169–170 and footnote (30) above. Also note Bourdieu's comments [1977, page 95, and footnote (7) above] on habitus and the capacities and limits influencing the invention of practices.

practice and structure, of the dialectics that are one with the temporally and spatially specific intersections between individual paths and the everyday workings of institutions. The same is also true of the origins of so-called authority constraints, frequently associated with dominant projects and other institutional projects. Moreover, we must not overlook the fact that the relative stringency of coupling constraints, the everyday ease or difficulty of reaching other particular projects immediately preceding or succeeding a new or redefined institutional project, also depends on the location and density of buildings and connecting routes deposited on the landscape through past social reproduction, through previous manifestations of the dialectics of practice and structure. It is therefore regrettable that all too many time-geography scholars have been content to ignore the underlying dialectics of their framework, to treat, implicitly or explicitly, coupling and authority constraints as if they were self-materializing givens, and so mislead or dissatisfy many of their readers.

An all too short note on language
Language is perhaps more tightly interwoven than anything else both with the specific intersections that occur between individual paths and institutional projects and with the dialectics that are inseparable from those intersections. Social reproduction and the details of everyday life, through which the individual and society are always becoming, are not possible without language and the signification, communication, and discourse it allows. Yet it would not be possible for language itself to be dynamic and always becoming without social reproduction and the details of everyday life.

  Language does not exist on its own. Its usage, whether in speaking, hearing, reading, or writing, always occurs in conjunction with institutionally and individually defined projects. Language is, in fact, "intrinsically involved with that which has to be done" (Giddens, 1979, page 4), with the day-to-day implementation and intergenerational transmission of social practices, conventions, and traditions, with the retention and interpersonal recounting of previously occurring actions and experiences (Wittgenstein, 1968). In one or more of these senses, language is fundamental to the bringing off of every individual-path–institutional-project intersection. Language is not only fundamental to the realization of such intersections because it is the medium through which the component tasks of projects are defined and made mutually understood; it is also fundamental because the external–internal and life-path–daily-path dialectics that lead the individual to specific intersections are dependent upon its acquisition and mental utilization. But an individual's acquisition of language—*or any other sign system*—and her internalization of its accompanying interpretative schema occur in adulthood as well as in childhood through specific instances of project participation, and thereby through the operation of her external–internal and life-path–daily-path dialectics[36]. In other words, (1) the vocabulary, and the phonemic, syntactic, semantic, and semiotic content of an individual's language, (2) her deep-seated sensitivity to the 'proper' social context for certain word choices, meanings, and colorations[37], (3) her "rhetorical devices, expressive effects, nuances of pronunciation, melody of intonation, registers of diction [and] forms of phraseology" (Bourdieu and Passeron, 1977, page 117), and (4) her ability to decipher, classify, interpret, and internally manipulate situations of varying complexity—all these things are continually the result of previous participation in social reproduction under spatially and temporally specific circumstances, and they are continually at the foundation of subsequent participation in social reproduction under spatially and temporally specific circumstances.

Because an individual's acquisition and internal usage of language cuts two ways, because her possession is both a prerequisite to and a consequence of intersections between her path and specific institutional projects, the converse of Wittgenstein's oft-cited observation on the imprisoning qualities of language also holds. It may well be true that "the limits of my language mean the limits of my world" (Wittgenstein, 1922); or, that the limits of what I can say and know through language mean the limits of the institutional projects I am able to participate in (and, hence, also the projects I can define for myself)[38]. But it is also true that the limits of my world mean the limits of my language; or, that the projects to which I am limited mean the limits of what I can say and know through language. Equally, just as one's words and wordless silences creates one's world, so does one's world create one's words and wordless silences (Olsson, 1980b; Silverman and Torode, 1980). To be a prisoner of one's language is to be a prisoner of one's own encounters with institutional practices in a given historical situation, a captive of one's own everyday contribution to social reproduction and transformation[39]. Therefore, language, which along with other systems of signs is easily equated with intention-affecting ideology and the sustenance of domination (Geertz,

[36] See footnote (23) above, on the persistence of socialization throughout an individual's life.

[37] That is, her use of different sociolinguistic speech variants (Bernstein, 1975a; Gouldner, 1976, pages 58–60).

[38] See earlier comments (pages 167–168) on the impact of institutional project participation on personally defined projects.

[39] See, for example, the following observation by Gouldner (1976, page 54): "The meaning of our imprisonment in language depends fundamentally on the fact that we are *not* just speakers, but that our languages are part of a larger life of practice, and vary with the nature of that practice. Those who largely live a passive contemplative existence, or others who relate to the world with sensuous aesthetic appreciation, and still others who view the world as an object to be acted upon, changed and used, all engage in fundamentally different forms of practice."

1964; Barthes, 1972; Gouldner, 1976; Olsson, 1978; Giddens, 1979), "can never express any *thing*" (Olsson, 1980b, page 11e)[40]. The employment of language in communication can only express the flow of society's structural relations and the internal relations of the individual, themselves derived from her unique accumulation of path intersections with the workings of society (Bally, 1965; Bourdieu and Passeron, 1977; Gregory, 1980a). Finally, these assertions regarding the dialectical interplay between language and what one does or does not practice become all the more pregnant if one is willing to acknowledge, in accordance with Lacan (1968) and others, that the rigid categorizations of a language are at the root of the repression which forms the unconscious, at the root of the taboo-laden, at the root of that which is absolutely forbidden to become a part of the individual's path (Olsson, 1980a)[41].

The transformation of language, like its constitution and reproduction, occurs primarily through institutional projects. Language is always *becoming* in the sense that its words, variable meanings, pronunciation, and grammar are either very gradually and unintentionally altered through daily use in stable and recurrent institutional projects, or are incrementally and sometimes radically changed through the introduction, abandonment, or modification of institutional projects and their associated path-coupling requirements. New or altered institutional projects bring new or altered interactional situations and contexts—configurations of people and objects which demand either that old words be given new meanings, or that completely new words and word–object relations come into being. By extension, since new or modified dominant institutional projects generate the most far-reaching changes in interactional situations and contexts, so they also foster the most dramatic linguistic changes, both in terms of simple word designations, as well as highly complex word-embedded symbolizations. Furthermore, the abandonment of institutional projects in general, and the elimination of dominant projects in particular, results either in the extinction of words and expressions within everyday usage, or in the disappearance of meanings, by removing the necessity of certain individual path conjunctions and thereby ending certain interactional situations and contexts.

[40] Ideology and ideology formation are concepts which carry with them a variety of connotations (Gouldner, 1976; Williams, 1977, pages 55–71, 109; Liedman, 1980b). Most of these connotations are intimately and complexly related in one way or another to social reproduction, structuration, and the dialectics associated with the time-geography of everyday life. However, the length and format of this essay do not permit consideration of these different connotations.
[41] To make such an acknowledgement is not necessarily to accept fully Lacan's structural interpretation of the emergence of the subject, which is well summarized in Coward and Ellis (1977, pages 93–121). For some critical reservations see Giddens (1979, pages 120–123).

Conclusions: anticlimactic variety
In this essay, I have stated many different things in the same way. The
same thing in many different ways. And, thus, this piece has repeatedly
turned in upon itself. This piece has also turned in on itself because while
my path has been caught up in the everyday project of preparing and
writing it, and while my path has been caught up in an effort to contribute
to current theories of social reproduction and structuration by integrating
them with time-geography, I have been inescapably subject to the same
dialectical relationships that I have tried to convey. This essay was also
designed to turn in on itself through you, and to turn on in you. For, even
if you have only just begun to understand what I have written, then its
impact on you should reach beyond the acquisition of some increased
awareness of social reproduction and the time-geography of everyday life
in general. You also should, on reflection, acquire at least some faint
comprehension of why it is that you are sitting here, right now, undertaking the project of reading these pages and reacting the way you are,
rather than doing something else elsewhere and reacting in other ways.
In other words, you should be able to see, at least dimly, some connection
between this particular moment and social reproduction and the time-
geography of everyday life in a given ongoing historical situation, between
this particular here and now and the dialectics of practice and structure,
between this particular here and now and your own external–internal
dialectics and life-path–daily-path dialectics, between the academic role
you are filling at this particular moment and all those other aspects of
your life and the workings of society in which they are enmeshed.
   Many other things could have been stated in this essay. Not only could
the language theme have been more fully developed, but much more could
have been said about many of the other grand categories of social theory—
power, ideology, class relationships, legitimation, alienation, patriarchy.
After all, each of these phenomena are both generated and expressed
through social reproduction and the dialectics associated with the time-
geography of everyday life, through the continual intersection of particular
individual paths with particular institutional projects at specific temporal
and spatial locations. The same is true of one of human geography's
grand themes—sense-of-place. But to have attempted to say more within
the confines of these pages would have been impractical and overly
ambitious in terms of the literature to be digested, synthesized critically,
and elaborated upon in the unfolding and discipline-transcending language
of time-geography.
   Many other things should have been stated in this piece. For example,
much more should have been stated about dominant institutional projects;
but perhaps these further elaborations would have been out of place, since
my concern here has been the theory of social reproduction and
structuration rather than the attributes of particular societal structures. In
addition, since reproduction subsumes the production and reproduction of

material goods, a truly adequate theory of social reproduction ought to encompass not only the detailed everyday dialectics of practice and structure, but also include some representation of the dialectical interplay between nature, or the physical landscape, and man and his practices (Ley, 1978; Gregory, 1980a). Another of the dialectics underlying time-geography, the path-convergence–path-divergence, or creation–destruction, dialectic (Pred, 1981a) lends itself, at least in part, to the capturing of the dialectical relationships between everyday practice and the transformation of nature[42]. But it also would have proved impractical and overly ambitious to deal with this additional theme within the limits of this essay.

Conclusion: climactic variety

There is a pathology to social reproduction and everyday life.

Society projects itself through the individual as the individual projects herself through society.

For you and me, for society as a whole, history and everyday life incessantly penetrate one another.

Acknowledgements. My deepest appreciation is due to Gunnar Olsson who—along with fine wines, good food, and Thursday afternoons—made the creation of this essay easier. It was written while I was receiving research support from the US National Science Foundation. It is reprinted, in modified form, with the permission of *Geografiska Annaler*.

References

Bally C, 1965 *La Language et la Vie* (Droz, Geneva)
Barthes R, 1972 *Mythologies* (Hill and Wang, New York)
Berger P L, Luckmann T, 1967 *The Social Construction of Reality: A Treatise in the Sociology of Knowledge* (Anchor Books, Garden City, NY)
Bernstein B, 1975a *Class, Codes and Control: Theoretical Studies Towards a Sociology of Language* (Schocken Books, New York)
Bernstein B, 1975b *Class, Codes and Control: Towards a Theory of Educational Transmissions* (Routledge and Kegan Paul, Henley-on-Thames, Oxon)
Bhaskar R, 1979a "On the possibility of social scientific knowledge and the limits of naturalism" in *Issues in Marxist Philosophy, 3. Epistemology, Science, Ideology* Eds J Mepham, D H Ruben (Harvester Press, Sussex) pp 107–139
Bhaskar R, 1979b *The Possibility of Naturalism: A Philosophical Critique of the Contemporary Human Sciences* (Harvester Press, Sussex)
Bourdieu P, 1977 *Outline of a Theory of Practice* (Cambridge University Press, Cambridge); originally published in 1972 as *Equisse d'une Théorie de la Pratique, Précédé de Trois Etudes d'Ethnologie Kabyle* (Droz, Geneva)
Bourdieu P, Passeron J C, 1977 *Reproduction in Education, Society and Culture* (Sage, London)
Bourdieu P, Passeron J C, 1979 *The Inheritors: French Students and Their Relation to Culture* (University of Chicago Press, Chicago)
Bourdillon M F C, 1978 "Knowing the world or hiding it: a response to Maurice Bloch" *Man* new series **13** 591–599

---

[42] The transformation-of-nature aspects of the creation–destruction dialectic are only hinted at in Pred (1981a).

Burawoy M, 1979 *Manufacturing Consent: Changes in the Labor Process Under Monopoly Capitalism* (University of Chicago Press, Chicago)

Carlstein T, 1981 *Time Resources, Society and Ecology: On the Capacity for Human Interaction in Space and Time* (Edward Arnold, London)

Cederlund K, 1977 "Administrativ verksamhet som projekt" *Svensk Geografisk Årsbok* **53** 81-92

Cooper L N, 1980 "Source and limits of human intellect" *Daedalus* Spring 1-17

Coward R, Ellis J, 1977 *Language and Materialism: Developments in Semiology and the Theory of the Subject* (Routledge and Kegan Paul, Henley-on-Thames, Oxon)

Cullen I G, 1978 "The treatment of time in the exploration of spatial behaviour" in *Timing Space and Spacing Time, 2. Human Activity and Time Geography* Eds T Carlstein, D Parkes, N Thrift (Edward Arnold, London) pp 27-38

Ellegård K, Hägerstrand T, Lenntorp ' 1977 "Activity organization and the generation of daily travel: two future alternatives" *Economic Geography* **53** 126-152

Elster J, 1978 *Logic and Society: Contradictions and Possible Worlds* (John Wiley, Chichester, Sussex)

Erikson E H, 1963 *Childhood and Society* second edition (W W Norton, New York)

Erikson E H, 1975 *Life History and the Historical Moment* (W W Norton, New York)

ERU, 1980 *Offentlig verksamhet och regional valfard: Perspektiv pa det politiska beslutsfattandet och den offentliga sektorn* Expertgruppen for forsnkning om regional utveckling (Statens offentliga utredningar, Stockholm) 6

Geertz C, 1964 "Ideology as a cultural system" in *Ideology and Discontent* Ed. D Apter (Free Press, New York) pp 42-67

Giddens A, 1973 *The Class Structure of the Advanced Societies* (Hutchinson, London)

Giddens A, 1976 *New Rules of Sociological Method* (Hutchinson, London)

Giddens A, 1979 *Central Problems in Social Theory: Action, Structure and Contradiction in Social Analysis* (University of California Press, Berkeley)

Gouldner A W, 1976 *The Dialectic of Ideology and Technology: The Origins, Grammar, and Future of Ideology* (Seabury Press, New York)

Gregory D, 1978 "The discourse of the past: phenomenology, structuralism, and historical geography" *Journal of Historical Geography* **4** 161-173

Gregory D, 1980a "The ideology of control: systems theory and geography" *Tijdschrift voor Economische en sociale Geografie* **71** 327-342

Gregory D, 1980b *Social Theory and Spatial Structure* (Hutchinson, London)

Hägerstrand T, 1970a "Tidsandvändning och omgivningsstruktur" in *Urbaniseringen i Sverige: En geografiska samhällsanalys* (Allmänna Förlaget, Stockholm) pp 4:1-4:146

Hägerstrand T, 1970b "What about people in regional science?" *Papers of the Regional Science Association* **24** 7-21

Hägerstrand T, 1972 "Tätortsgrupper som regionsamhällen: Tillgången till förvarvsarbete och tjänster utanför de storre staderna" in ERU (Expertgruppen for regional utredningsverksamhet) *Regioner atta leva i* (Allmänna Förlaget, Stockholm) pp 141-173

Hägerstrand T, 1974a "On socio-technical ecology and the study of innovations" *Etnologica Europaea* **7** 17-34

Hägerstrand T, 1974b "Tidsgeografisk beskrivning: Syfte och postulat" *Svensk Geografisk Årsbok* **50** 86-94

Hägerstrand T, 1975 "Space, time and human conditions" in *Dynamic Allocation of Urban Space* Eds A Karlqvist, L Lundquist, F Snickars (Lexington Books, D C Heath, Lexington, Mass)

Hägerstrand T, 1976 "Geography and the study of interaction between nature and society" *Geoforum* **7** 329-334

Hägerstrand T, 1977 "On the survival of the cultural heritage" *Ethnologica Scandinavica*
1 7-12

Hägerstrand T, 1978 "Survival and arena: on the life history of individuals in relation
to their geographical environment" in *Timing Space and Spacing Time, 2.*
Human Activity and Time Geography Eds T Carlstein, D Parkes, N Thrift (Edward Arnold,
London) pp 122-145

Hawkes T, 1977 *Structuralism and Semiotics* (University of California Press, Berkeley)

Hoppe G, 1978 *Formal Education and Life-Path Development* Department of Human
Geography, Kulturgeografiskt Seminarium

Kosík K, 1978 *Det konkretas dialektik: En studie i människans och världens
problematik* (Röda Bokförlaget AB, Göteborg); originally published in 1963 as
Dialektika konkrétniko; also available in English as *Dialectics of the Concrete*
Boston Studies in the Philosophy of Science, 52 (1976)

Lacan J, 1968 *The Language of the Self: The Function of Language in Psychoanalysis* (Johns Hopkins University Press, Baltimore)

Lenntorp B, 1976 *Paths in Space-Time Environments: A Time-Geographic Study of
Movement Possibilities of Individuals* Lund Studies in Geography, Series B, 44
(Gleerup, Lund)

Lenntorp B, 1978 "A time-geographic simulation model of individual activity
programmes" in *Timing Space and Spacing Time, 2. Human Activity and Time
Geography* Eds T Carlstein, D Parkes, N Thrift (Edward Arnold, London)
pp 162-180

Ley D, 1978 "Social geography and social action" in *Humanistic Geography:
Prospects and Problems* Eds D Ley, M Samuels (Maaroufa, Chicago)

Liedman S E, 1980a "Det är i vardagen allt stort sker" *Dagens Nyheter* 4

Liedman S E, 1980b *Surdeg: En personlig bok om idéer och ideologier* (Författar-
förlaget, Stockholm)

Lukács G, 1976 *History and Class Consciousness* (MIT Press, Cambridge, Mass);
originally published in 1923 as *Geschichte und Klassenbewusstein* (Der Malik-
verlag, Berlin)

Mårtensson S, 1978 "Time allocation and daily living conditions: comparing regions"
in *Timing Space and Spacing Time, 2. Human Activity and Time Geography*
Eds T Carlstein, D Parkes, N Thrift (Edward Arnold, London) pp 181-197

Mårtensson S, 1979 *On the Formation of Biographies in Space-Time Environments*
Lund Studies in Geography, Series B, 47 (Gleerup, Lund)

Mead G H, 1934 *Mind, Self and Society* (University of Chicago Press, Chicago)

Olander L O, Carlstein T, 1978 "The study of activities in the quaternary sector" in
Timing Space and Spacing Time, 2. Human Activity and Time Geography
Eds T Carlstein, D Parkes, N Thrift (Edward Arnold, London) pp 198-213

Olsson G, 1978 "On the mythology of the negative exponential or on power as a
game of ontological transformations" *Geografiska Annaler* **60** Series B, 116-123;
reprinted in *Birds in Egg/Eggs in Bird* (Pion, London) 1980, pp 30e-39e

Olsson G, 1980a "Anmälan av Erik Wallin's avhandling 'Vardagslivets generativa
grammatik'" *Svensk Geografisk Årsbok* **56** 89-94

Olsson G, 1980b "Hitting your head against the ceiling of language" in *Birds in
Egg/Eggs in Bird* (Pion, London) pp 3e-18e

Olsson G, 1980c "Toward a mandala of thought-and-action" in *Birds in Egg/Eggs in
Bird* (Pion, London) pp 19e-29e

Olsson G, 1981 "Toward a sermon of modernity" in *Geography as a Spatial Science:
Recollection of a Revolution* Eds M Billinge, D Gregory, R Martin (Macmillan,
London) forthcoming

Parkes D, Thrift N, 1979 "Time spacemakers and entrainment" *Transactions of the
Institute of British Geographers* new series 4 353-371

Parkes D, Thrift N, 1980 *Times, Spaces, and Places: A Chronogeographic Perspective* (John Wiley, Chichester, Sussex)

Parsons T, 1949 *The Structure of Social Action* (Free Press, New York)

Peet R, 1975 "Inequality and poverty: a Marxist-geographic theory" *Annals of the Association of American Geographers* **65** 564–571

Persson C, 1977 "Omgivningsbilder och deras roll i beslutsprocessen: Några reflektioner kring teori och metodik" *Svenska Geografisk Årsbok* **53** 30–43

Piaget J, 1955 *The Child's Construction of Reality* (Routledge and Kegan Paul, Henley-on-Thames, Oxon)

Popper K, 1966 *The Open Society and Its Enemies* 2 volumes (Routledge and Kegan Paul, Henley-on-Thames, Oxon)

Poster M, 1978 *Critical Theory of the Family* (Seabury Press, New York)

Pred A, 1973 "Urbanisation, domestic planning problems and Swedish geographic research" *Progress in Geography* **5** 1–76

Pred A, 1977 "The choreography of existence: comments on Hägerstrand's time-geography and its usefulness" *Economic Geography* **53** 207–221

Pred A, 1978 "The impact of technological and institutional innovations on life content: some time-geographic observations" *Geographical Analysis* **10** 345–372

Pred A, 1981a "Of paths and projects: individual behavior and its societal context" in *Behavioral Geography Revisited* Eds R Golledge, K Cox (Methuen, London) pp 231–255

Pred A, 1981b "Production, family, and 'free-time' projects: a time-geographic perspective on the individual and societal change in nineteenth-century U.S. cities" *Journal of Historical Geography* **7** 3–36

Radcliffe-Brown R, 1940 "On social structure" *Journal of the Royal Anthropological Institute* **70** 1–12

Richards M P, 1974 *The Integration of a Child into a Social World* (Cambridge University Press, Cambridge)

Schutz A, 1967 *The Phenomenology of the Social World* (Northwestern University Press, Evanston, Ill.)

Silverman D, Torode B, 1980 *The Material Word* (Routledge and Kegan Paul, Henley-on-Thames, Oxon)

Strenski I, 1974 "Falsifying deep structure" *Man* new series **9** 571–583

Thompson E P, 1967 "Time, work-discipline, and industrial capitalism" *Past and Present* **38** 56–97

Thrift N, 1977 "Time and theory in human geography: Part II" *Progress in Human Geography* **1** 413–457

Thrift N, 1979 *The Limits to Knowledge in Social Theory: Towards a Theory of Practice* Australian National University, Department of Human Geography, Canberra

Thrift N, 1980 "Local history: a review essay" *Environment and Planning A* **12** 855–862

Thrift N, 1982a "Social theory and human geography" *Area* **13** (forthcoming)

Thrift N, 1982b "Owners' time and own time: a geography of capitalist time consciousness, 1300–1880" in *Space and Time in Geography: Essays dedicated to Torsten Hägerstrand* Lund Studies in Geography, Series B, 47 (Gleerup, Lund)

Thrift N, Pred A, 1981 "Time-geography: a new beginning (a reply to Alan Baker's historical geography: a new beginning)" *Progress in Human Geography* **5** 277–286

Törnqvist G, 1979 *On Fragmentation and Coherence in Regional Research* Lund Studies in Geography, Series B, 45 (Gleerup, Lund)

Touraine A, 1977 *The Self-Production of Society* (University of Chicago Press, Chicago)

Wallin E, 1974 "Yrkeskarriär och stabilitet i bosättningen" in ERU (Expertgruppen for regional utredningsverksamhet) *Ortsbundna levenadsvillkor* (Statens offentliga utredningar, Stockholm) 2 298-322

Wallin E, 1980 *Vardagslivets generativa grammatik—vid gränsen mellan natur och kultur* Meddelanden från Lunds Universitets Geografiska Institution avhandlingar LXXXV, Lund

Weintraub K J, 1978 *The Value of the Individual: Self and Circumstance in Autobiography* (University of Chicago Press, Chicago)

Williams R, 1961 *The Long Revolution* (Chatto and Windus, London)

Williams R, 1973 *The Country and the City* (Chatto and Windus, London)

Williams R, 1977 *Marxism and Literature* (Oxford University Press, Oxford)

Williams R, 1979 *Politics and Letters: Interviews with New Left Review* (New Left Books, London)

Wittgenstein L, 1922 *Tractatus Logico-Philosophicus* (Routledge and Kegan Paul, Henley-on-Thames, Oxon)

Wittgenstein L, 1968 *Philosophical Investigations* third edition (Macmillan, New York)

Solid geometry: notes on the recovery of spatial structure[†]

Derek Gregory

"A capacity for orderly drowning. In a schoolboy craze ... in ... maps. Concentrations so knotted that they left the world to one side."
George Steiner, *The Portage to San Christobal of A. H.*

Introduction

In this essay I shall try to explore some of the relations between modern human geography and modern social theory through an examination of the concept of spatial structure. The exercise is necessarily a preliminary one, but in order to provide some sort of context for what follows, the first section retraces in outline the emergence of a spatial tradition in geography. In this first part of the essay I argue that conventional historiographies have frequently obscured two important moments in that early movement. In both cases continuities have been fractured and dislocations sharpened. In particular, I shall claim that the debate over exceptionalism in the 1950s—that is, Schaefer's insistence that geography break with the classical Hartshornian orthodoxy and search for 'morphological laws'—was a mischievous one which effectively proscribed a necessary dialogue between two closely connected discourses. I shall also suggest that the resonances of that *débâcle* proved to be especially formative in that they eventually shattered the interrogation of the frozen spatial lattice, which had characterised so much of the *oeuvre* of the locational school, once it became clear that its technical procedures and their derivative point-process models were implicated in a structuralism (Olsson's "dialectics of spatial analysis") which posed a series of acutely difficult questions about the status of the purely geometric constructs which had dominated the discursive formation of postwar human geography. Their resolution demanded an engagement with social theory and social philosophy.

What few appeared to realise at the time was that the foundations of the locational school had been laid by workers who were themselves no strangers to social theory and philosophy. The apparently motionless rings of von Thünen's model of agricultural land use scythed their way out into a wider discussion of the historical relations between landlords and peasants; the isodapanes which circumscribed Weber's model of industrial location were set spinning in his subsequent discussion of the dynamics of capitalist industrialisation, and they then spiraled through a programmatic

† The title is taken from a short story by Ian McEwan in *First Love, Last Rites*: its sense of solid geometry both conveys the concentrations and the contortions to which the elusive concept of spatial structure has been subjected.

cultural sociology which roundly rejected the possibility of an autonomous location theory; and the ring-mosaics of the Chicago models of urban land use were not merely cages for the contemplation of a brute and invariant 'natural order' through the unyielding *grille* of human ecology, but also springs for the explication of the historical formation of a definite 'moral order' founded on public discourse and communication. In much the same way that Durkheim had tried to secure a vacant space for sociology by treating society as a reality *sui generis* (explicable in terms of itself and as such detached from *la géographie humaine*), so through what Sack has described as the 'spatial separatism' of the locational school the task of a reconstituted human geography came to be defined in equally exclusive notations—namely, as the explanation of spatial structures by the determination of spatial processes. As Blaut recognised, this is essentially a Kantian dualism, but I want to suggest that its resolution does *not* depend upon a closure of the locational school *tout court*. On the contrary, we still have much to learn from a considered examination of its early texts and from a careful reformulation of the concepts of spatial structure which they contained.

It is necessary to say all this because the contemporary union with social theory has paradoxically been consummated (with some notable exceptions) by a premature withdrawal from the analysis of spatial structure. I shall try to show in the second section that the so-called structuralist critique—drawing much of its inspiration from Althusser's symptomatic reading of Marx—equates all forms of spatial analysis with empiricism, and hence represents space as an 'expression', a 'realisation', or a single 'aspect' of an ensemble of social relations which together constitute the only authentic object of a properly scientific study. In the light of this critique, conventional spatial analysis characteristically conflates pattern and structure, and it is necessarily incapable of providing theoretical categories sharp enough to dissect spatial surfaces and so disclose their generative social structures. Coincidentally, a voluntarist critique—which derives from interpretative sociology and constitutive phenomenology—has registered a series of objections to all concepts of structural determination, both social and spatial, and hence identifies spatial analysis with a 'geometric determinism' incapable of recognising and specifying the subjective constitution and social signification of conceptions of space. In both cases, spatial analysis has been overwritten by a social analysis that has often taken an extremely abstract form.

There are clear and important differences between these two schematic versions of social theory, and I have indicated elsewhere (Gregory, 1981b) that I regard both of them as deficient. Nevertheless, in the third section of this essay I shall suggest that where they succeed is in establishing, in strikingly opposite ways, social reproduction as a moment in the formation of spatial structures. Although this constitutes a vital advance, I shall also note that they fail to complete the cycle, since they do not identify the formation of spatial structure as a moment in social reproduction. These twin claims obviously oversimplify what should properly be regarded as highly unstable and much more nuanced formulations, but I think they are still robust enough to stake out a lacuna which ensnares concepts both of spatial structure and social reproduction. To clarify: neither scheme provides an adequate understanding of situated social practice, and as a result I shall argue that neither of them captures the recursive motion of human history *as it is instantiated in human geography*. This last clause is neither redundant nor residual, but in case it should be misunderstood let me state clearly that it is *not* intended to advance what Soja has called 'geographical materialism' (an uneasy dual of historical materialism), and still less is it intended to defend a distinctive discipline founded on spatial analysis. Rather, my purpose is to underscore the importance of concepts of spatial structure to the human and social sciences as a whole. Even so, it is impossible to move very far into the debate without confronting the efforts of previous writers to demarcate just such an exclusive domain— an 'uncommon ground', roped off, guarded, and under constant surveillance. I shall obviously be obliged to acknowledge materials of this sort, but I shall do so only to disentangle the concepts of spatial structure which they entail, and I wish to avoid complicity in their exceptionalist projects.

Spatial formalism

Postwar geography was dominated by discussions of Hartshorne's *The Nature of Geography*, first published in 1939. His notion of a chorological science was predicated on a discussion of concepts of spatial structure, and although these were not developed in any systematic and consistent fashion (there are a number of reasons for this), it is possible to caricature some basic elements. The "very essence of geographic thought", according to Hartshorne, was "integration of phenomena in spatial associations", and although he did not consider that these could disclose any generative processes, their dissection was not simply limited to a morphology. "The significance of patterns depends entirely on the extent to which they depict significant relations in the location of different places in reference to each other", he wrote, and in his "theoretical approach to regional geography" these spatial integrations were represented as complex, multi-dimensional surfaces, displayed within spatial coordinate systems, which could, at least in principle, be expressed as functional equations. He regarded "the purely geometric factor of location" as "more distinctively geographic than any other", and, since his exegesis was largely confined to a German intellectual tradition, he was keenly aware of location theory and especially the models of von Thünen and Weber. It is true that he regarded their explanations as deficient, but such geometries were "ever essential for a geographic interpretation" as they were capable both of a *formal* and of a *functional* inflection (Hartshorne, 1939, pages 438–439). This is not the Hartshorne who haunts conventional historiographies.

In its fundamentals—which I take to exclude the entirely trivial discussion of 'idiographic' and 'nomothetic' knowledges—*this prescription is consonant with Schaefer's programmatic spatial science.* To be sure, Schaefer's concept of spatial structure was even more starkly geometric (and in an important sense therefore even more exceptionalist) than Hartshorne's, but its exclusions were registered in similar terms. For example, he complained that geographers "do not always clearly distinguish between, say, social relations on the one hand and spatial relations .. on the other", and he insisted that "spatial relations are the ones that matter in geography, and no others". And although his belief that geography was "compelled to produce morphological laws" was phrased in a lexicon which Hartshorne explicitly rejected, his advocacy of a spatial analysis that contained "no reference to time or change" was uncomfortably close to the orthodoxy he was supposed to be overturning. "This is not to deny that the spatial structures we explore are, like all structures anywhere, the result of process", he conceded, but—like Hartshorne—"the geographer, for the most part, deals with them as he finds them, ready-made" (Schaefer, 1953, pages 243–244).

Although Bunge compared the Hartshorne–Schaefer *Methodenstreit* to the dispute between Hegel and Feuerbach, it is extraordinarily difficult to reconcile this with the emphasis that both placed on the geometry of landscape and their common exclusion of processes from concepts of spatial structure. Indeed, in an angry reply Hartshorne spoke of an "essential agreement" between himself and Schaefer, and maintained that "the picture subsequently constructed of major disagreement is therefore false" (Hartshorne, 1955).

This undercurrent of what, in accordance with Sack (1974), I have called spatial separatism surged to the surface during the 1960s, flowing over what Harvey later described as "the special bridges between geography and the formal spatial languages that constitute the whole of geometry" (Harvey, 1969, page 227). Its course was unwavering. Even when Bunge widened Schaefer's spatial relations to include 'spatial processes' (movements) and 'spatial structures' (forms), his suggestion that the latter could be defined "most sharply by interpreting 'structure' as 'geometrical' " ensured that "the science of space [would find] the logic of space a sharp tool" (Bunge, 1962, pages 211–212). This sort of intellectual surgery always had its critics, of course, but three years later Haggett's *Locational Analysis in Human Geography* effectively silenced the antivivisection lobby by recalling that "the geometrical tradition was basic to the original Greek conception of the subject". "Many of the more successful attempts at geographical models", he continued, had their origins "in this type of analysis" (Haggett, 1965, page 15). There is surely no need here to review the steady stream of work on point-patterns and point-process models, on network structures and taxonomies of spatial hierarchies, and on the topology of space-filling and space-partitioning processes which followed.

But it is important to notice that much of this was conveyed through the most elementary of geometries, slicing through low-dimensional spaces to uncover the simplest of their symmetries. Perhaps we should recall Lévi-Strauss's defiant definition: "scientific explanation consists not in moving from the complex to the simple, but in the replacement of a less intelligible complexity by one which is more so" (Lévi-Strauss, 1966, page 248). From this perspective, the first-generation models of spatial structure were at once simple and unintelligible. The criticism that they were 'unrealistic', 'inapplicable' or whatever was (or rather, could have been) much more than a profoundly conservative rejection of an alien philosophy. It is not difficult to explicate such objections and to high-light their palpable sense of outrage at "so much missing": the models were concerned with *reduction*, with the collapse of spatial structures into coarsely-textured, low-level surfaces, rather than with '*de-construction*', their expansion into richly-textured, high-level structural domains.

This is to translate spatial formalism into structuralism, and I must now substantiate such a movement. This is a difficult task because the translation is constantly changing, as I will show, and it is hard to be sure of its direction. But we can begin with a remarkably early essay in which Blaut (1961) dismissed simple spatial formalism because "every empirical concept of space must be reducible by a chain of definitions to a concept of process". In traditions like Schaefer's by contrast:

"Structure and process, form and matter, being and becoming—each of these ancient metaphysical oppositions is woven into the contrast between spatial morphology and the content of space, between, on the one hand, a kind of permanent and rigid spatial framework, distinct from the material world yet providing it with shape and pattern, purely geometric yet somehow registering on maps, and on the other hand the flux of changing process." (Blaut, 1961, page 4)

Put somewhat differently, Lévi-Strauss has argued that structuralism depends on these elisions being made through an examination of historical sequences, however disjointed and imperfect they may be, because it is history alone which "makes it possible to abstract the structure which underlies the many manifestations and remains permanent through a succession of events" (Lévi-Strauss, 1969, page 21; see also Gregory, 1978a). In much the same way, Cliff and Ord (1981) have recognised that the disclosure of spatial structures demands temporal collation, so that they scan a sequence of maps for systematic autocorrelation structures in space and in time in order to specify 'reactive' and 'interactive' processes. In a sense, therefore, albeit a very special one, structure and process become dissolved into one. We can press this still further. In a lucid critique, Sack (1972) complained that within a geometric tradition (which this still is) "geography becomes the analysis of points and lines without regard to what these symbols represent". The key term is 'symbol', however, for structuralism displaces (literally deconstructs) the

meaning-endowing subject and is not concerned with "what these symbols
represent" (the signifieds) at all, but instead with the structures composed
by the signifiers, here the "points and lines". And it is for exactly this
reason that I suggest spatial formalism is susceptible to a structuralist
transformation. Both procedures entail moving 'beneath' the conscious
designs of purposive human subjects in order to expose an essential,
unacknowledged logic which binds them together in stable structures.
And in precisely this context Gould (1974, page 12) has reminded us
that "what is surprising is not the uniqueness of patterns of spatial
organisation ... but their extraordinary similarity. The constraints may
vary from place to place, but on their release familiar, consistent and non-
surprising patterns evolve. There are, perhaps, deep structures of human
behaviour underlying these repetitive patterns, if only we have knowledge
enough and the eyes to see."

    Yet the status of these structures remains problematic, which is why I
believe their recovery has necessarily involved a 'linguistic turn' in modern
human geography, comparable with that described by modern critical
theory, and marked by a parallel, deep concern with linguistic philosophy
and the examination of discourse. This owes much to post-Wittgensteinian
linguistic philosophy and, as such, to structuralisms which depart from the
restricted ambit of Lévi-Strauss's project—to which objections can most
certainly be made.

    I want to consider two of these explorations here, exemplified by the
writings of Gunnar Olsson (1980) and Peter Gould (1980), and to tease
out two threads of considerable importance. *First*, both of them are
impatient with the limited range of languages that have been deployed in
traditional social science, because these correspondingly limit our visions
of and knowledges about the social world: that is, our ability to disclose
those 'deep structures' is contained within, and is given by, the structure
of our language. For the most part, they argue, we have paced around
one small cell in the prisonhouse of language, backwards and forwards,
dismally failing to extend the horizon visible through its rigid bars. More
particularly, Olsson claims that the categories of conventional spatial
analysis are unable to mirror the flow and pulse of human agency, which
is "inherently so ambiguous that it will resist any attempt to catch it in
the firm categories of formal reasoning", and Gould notes that those same
geometric categories "crush", "diminish", and "destroy" the "multidimensional spaces" (structures) in which we live. Their separate
involvements with surrealism and with polyhedral dynamics are not,
therefore, antinomies, but derive instead from a conjoint concern with
languages capable of expressing the 'internal relations' or the 'structural
tissue' of the social world. And in neither case does this resolve into a
metaphysical idealism. On the contrary, as Rubinstein's (1981) marvellous
discussion of the connections between Marx and Wittgenstein has shown
more generally, concerns of this sort can explicate a concept of social

practice which transcends the exhausted duality between subjectivism and objectivism. This is latent within both of these schemes so that, *second*, therefore, both of them provide glimpses into a realm of what we might think of as 'interactive structures'. Olsson (1980) suggests that these are opaque to conventional spatial analysis because "With its traditional stress on space, measurability and visual landscape, geography has committed itself to the surface features of the external. Since the external is in things rather than relations, we have produced studies of reifications in which man, woman, and child inevitably are treated as things and not as the sensitive, constantly evolving beings we are" (page 47e). Such discourses are thereby embedded in structures of domination and their transcendence, in turn, requires a critique of ideology. Its task will be to disclose what Olsson calls the "ontological transformations in which things turn into relations and relations into things, visible into invisible, invisible into visible"; and while these are written "in the signs", effected through the medium of language, they are not purely linguistic, intra-discursive reversals because (to oversimplify) "internal relations are social relations, and ... social relations are internal relations" (page 10e). "Geographic space is a prison" to Olsson (page 197b), partly through its unexamined translation into a geometric conception of spatial structure, but partly too because these categories inform, define, and legitimate an extradiscursive social practice which leaves little room for "the constant groping of autonomous man" (page 204b). It follows that a new social science will have to be realised through a language capable of conveying "the creativity inherent in the human condition [which] gets its nourishment exactly from a balanced and dialectical interplay between the two forces of society and individual, of public and private, of macrocosm and microcosm" (page 34e).

These recursive movements between agency and structure reappear in polyhedral dynamics. Gould is convinced that they can never be humanely confined within the restricted frameworks of conventional spatial analysis, and to the extent that its propositions routinely feed into and flow from specific social practices then an essential moment in reconstruction is, once again, the critique of ideology. This will allow, via an appropriate language of structure written in algebraic topology, the recovery of a conception of structure as a multidimensional space, "a backcloth against which existence and action can, or cannot, take place", and the description of transformations both in "backcloth" and in "traffic" as ordered sets of relations anchored in the emergent connectivity of the structure itself. This is formally equivalent, I think, to a recognition that structures are both *conditions* and *consequences* of action. Obviously the symmetries involved are no longer simple, but their careful disclosure is intrinsically "emancipatory": "To be aware of one's self, in a geometry that forbids and allows, induces emancipatory self-reflection that raises the question of how the geometries might be changed" (Gould, 1982).

These two schemes can be connected up to critical theory easily enough, and their emphases on discursive formations, their engagements with structuralism, and the energising impulse of critique all have their resonances in the work of the Frankfurt School—in particular in the more recent writings of Habermas (1979). But they also expose the stark 'one-dimensionality' of conventional conceptions of spatial structure, and these echoes of Marcuse are by no means accidental. Robson (1973, page 90) once prefaced an analysis of spatial patterns with Pascal's awesome admission, "the silence of these eternal spaces terrifies me", and we should now be able to understand what it is that the structures, which (as it were) 'speak through' spatial analysis, are so frighteningly silent *about*: the transformations which, according to Rilke, our life passes *in*. Discursive formations are not, cannot be, 'innocent', and the restricted dimensionality of conventional spatial analysis implicates spatial science (whether we like it or not) in the production and reproduction of highly distorted structures of domination. It does so precisely because it represents spatial structures as flat, frozen lattices, the most solid of all geometries, massive and immobile. Breaking out from this planar prison means *enlarging* rather than *erasing* our conceptions of spatial structure: otherwise we are condemned to be mute spectators at Harvey's contest between "the dialectics of the social process and the static geometry of the spatial form" (Harvey, 1973, page 307).

Social structure, social practice, and spatial formation

We can maintain a thread of continuity with the preceding materials, and at the same time prepare for the succeeding discussion, by unpicking some of the strands of social theory which have been imperfectly spliced into time-geography and structural Marxism, and by disclosing the uneven texture of their theoretical fabrics. There are obvious differences between the two projects, but taken together they describe parallel paths from social structure through social practice to spatial formation, and yet both conspicuously fail to elucidate the return relations which are essential moments in social reproduction and historical transformation. In what follows they are not called out simply to be dismissed, because, as Giddens (1979a) and Habermas (1979) have shown, a critical science must concede the force of the structuralist critique, and these two formulations certainly register important advances over that of Lévi-Strauss.

Time-geography shares the concern with language which was identified earlier, with the development of what Hägerstrand (1973, page 77) describes as a "new notation", something like "the score of the composer", which can record the continuous flow of practical life as it unfolds in space and time—what Pred (1977) has termed the "choreography of existence". Hägerstrand believes (page 80) that this will eventually be capable of mathematical expression, and even speculates on the development of "a kind of applied topology, perhaps like laying a jig-saw puzzle with rubber pieces", although one of his co-workers suspects that "there is no form of mathematics available so far which does good justice to the time–space logic of social and ecological systems in the way that there already was a convenient form of mathematics available for explaining Einsteinian relativity theory in physics" (Carlstein, 1980, page 302). In fact, both Harvey (1969) and Rose (1977) have equated Hägerstrand's original conception with the analytic geometry developed by Minkowski for exactly that purpose. But what is remarkable about these phrasings is the way they mimic Gould's with such fidelity; and it should also be clear that they too demand that the geometric tradition "should not be abandoned but strengthened". Again, this takes the form of a move to a higher dimensionality. Although some commentators have concluded that time-geography is committed to a structural functionalism which cannot reveal "underlying structures in society" (Jensen-Butler, 1981, page 47), its models have in fact been translated into an explicitly structuralist problematic. To some degree, I think, this was implied by Hägerstrand's (1978, pages 129, 144) early sketches of "the conditions which circumscribe man's actions" and of "the relationship between the space–time trajectories of people and the underlying structure of options", understood as the ensemble of "choices and combinations ... made available by the 'geographies' ... that people have come to be part of". But these programmatic notes can be read in a number of different ways, and here I will confine my comments to Carlstein's writings which, because they are more developed, are less equivocal.

Carlstein draws upon the classical Saussurian distinction between *langue* (the repertoire of the possible) and *parole* (the realisation of the concrete), which played such a formative role in Lévi-Strauss's first productions, to define structure as "the result of the aggregation and interaction of constraints, whereby some solutions are feasible and others are not" (Carlstein, 1980, page 58). These constraints are the capability, coupling, and authority (or steering) constraints derived directly from Hägerstrand's original model, and this makes them insecure—even arbitrary—because if the scheme is to be internally consistent then the identification of the structures which the constraints assume clearly cannot rest on a casual empiricism, but needs to be bound into a properly constituted social theory. Hägerstrand's remarks about the importance of power are little more than gestural in this context, and it is the failure to provide a more rigorous discussion which opens time-geography to such variant (and discrepant) readings. Carlstein is especially sensitive to the charge that time-geography can only recover what Goddard (1972) terms more generally 'pseudostructures', control functions which ensure conformity with the normative order, rather than any 'deeper' structures which in some less contingent way allow for its contestation. The distance between the two is that between synchrony and diachrony. To close the space, Carlstein argues that the effects of the various constraints will not be

additive, and that a concept of structural causality is required to determine
a "hierarchy of reciprocal limitations" between them. The purpose of
this manoeuvre is to avoid being inculpated in the failings of functionalism
(although, as I will show, this is unsuccessful) by specifying the "limits of
functional compatibility" of different structures in such a way that
contradictions *within* structures can be distinguished from contradictions
between structures, so that both 'variations' and 'transformations' can be
accounted for within the same problematic. These differentiations
tremble on the edge of a structural Marxism, and Carlstein is scarcely the
first to make use of them (Godelier, 1972; Harvey, 1973, pages 291–292;
Friedman, 1974), but although he acknowledged his debt to the
anthropologies of Friedman and Godelier, these borrowings are never
articulated in any systematic way. Instead, Carlstein uses them to
delineate "possibility boundaries" in space–time, whose prisms correspond
to (or *map out*) the underlying structure. Insofar as this is supposed both
to have 'negative' and 'positive' functions—to "both forbid and allow"—
then webs of space–time interactions can be represented as the realisations
of a generative social structure. It follows from this that the repetition of
characteristic paths in space and time, the 'freezing' of a spatial structure
as individuals are routinely directed along particular pathways, can be
seen as the product of an invariant structural template, and that, co-
equally, any structural transformation ought to be mirrored in a dislocating
sequence of movements in space and time (Carlstein, 1980, page 47).

Yet the constitution of these structures and their transformations
remains elusive, and Jensen-Butler (1981, page 57) has called for a more
inclusively materialist conception which can explain "changes in the nature
of the constraints in the time–space model". This is not the straight-
forward matter he assumes it to be, however, and if we follow structural
Marxism more closely we can begin to sharpen these images and at the
same time detect some deficiencies in their composition. This is a delicate
task, made all the more difficult because structural Marxism is far from a
unitary discourse. But some of its most important elements have their
origins in the series of 'symptomatic readings' of Marx's texts prepared by
Althusser and his collaborators at the École Normale Supérieure in the
mid-1960s, and taken together these provide a *grille* of concepts which
can be fitted over the preceding discussions (Althusser and Balibar, 1970).
In a sense, of course, Althusser's project was again indicative of a deep
concern with language, and more particularly with the need to break
from the chronic reproduction of social forms as unexceptional categories
in an unexamined discourse: that is, from ideology. This is not the
place to spell out the rationalist epistemology which Althusser proposed
as a guarantee of scientific status, still less to list the indictments which
have been made of it; but many of these have been framed in terms of
a critique of structuralism which connects back to its linguistic basis.

And whereas most of these objections are decisive, there are two couplets which can be rescued and reformulated. These are the relations, first, between *langue* and *parole* (which Saussure mistakenly considered as an antinomy rather than a dialectic), and second, between *présence* and *absence* (which Saussure also presented in oppositional form, but which Derrida re-presents in dialectical notation). The two are closely connected, and the discussion which follows separates them purely for analytical convenience.

One of Althusser's central concepts is that of *social formation*. This was defined as a combination of modes or submodes of production structured by a dominant mode of production, which is itself structured by its differentiation into three 'relatively autonomous' economic, political, and ideological levels, governed by the 'matrix role' of the economy which determines the relations between these levels. This means that it must be possible to specify the conditions which are necessary for the realisation of a particular combination: or, to put it somewhat differently, "the conditions of existence of a particular mode of production prescribe the limits of variation in the structures of the economic, political and ideological levels if the mode of production is to survive" (Althusser and Balibar, 1970; Cutler et al, 1977). This speaks directly to Carlstein's project (although Althusser and Godelier are not saying exactly the same thing), but in fact he rejects the concepts of mode of production and social formation (1980, page 249) and, for reasons which I will discuss shortly, substitutes the underidentified 'activity system'.

Nevertheless, these formulations have been important in elucidating concepts of spatial structure, and Santos (1977, page 5) has drawn upon the heterogeneous traditions from which they derive to claim that "the mode of production amounts to only a possibility to be realized, and the [social] formation ... [to] the realized possibility". This is much closer to structuralism than Althusser would allow—he had repeatedly insisted that his *combination* is not the structuralist *combinatoire*—but Santos maps the one onto the other in order to explicate what, in accordance with Mathieu, he terms the 'spatial formation'. He argues that "taken individually, each geographic form is representative of a mode of production or of one of its moments", so that these elements of spatial structure "constitute a language [*langue*] of the modes of production". However, when taken together, their combination is effected through the realisation of a social formation which, because it can only occur "*in* space and *through* space", discloses what Soja (1980) has called a "full and equally salient spatial homology" which is therefore functionally equivalent to *parole*. These distinctions enable Santos to treat the history of modes of production as a *succession* of spatial forms, and the history of social formations as a *superimposition* of those successions, and in doing so to move along paradigmatic and syntagmatic axes to conclude that "modes of production write history in time; social formations write it in space".

This is more than an empty aphorism, although it is far from acceptable in its present form, and it anticipates the three-fold sense of *difference* entailed by modern theorems of structuration (Giddens, 1979a). Indeed, Vilar (1973) was making much the same point, in a different vocabulary, when he reminded Althusser that "history is not only an interlacing of times, but of spaces as well". Althusser's examination of various concepts of historical time in classical economics had prompted him to assign differential 'temporalities' to each level of the social formation. Since this is not an expressive (Hegelian) totality, he argued that "it is no longer possible to think the process of development of the different levels of the whole in the same historical time"; for example, the economic level of the capitalist mode of production contains "different rhythms which punctuate the different operations of production, circulation and distribution", and the concepts of these temporalities therefore have to be constructed "out of the concepts of these different operations" (Althusser and Balibar, 1970, pages 99–101). This logic is easily extended to meet Vilar's objection, and Lipietz (1977) has in fact suggested that concepts of 'spatialities' ought to be constructed in exactly the same way, so that a concept of spatial structure flows from a concept of social structure, as "a correspondence between presence/absence (in space) and participation/ exclusion (in the [social formation])". Although these can also be seen as correspondences between the distribution of 'places' in spatial structure, and the distribution of 'places' in social structure, they cannot be represented as vectors in a single space. Althusser's declaration that there is no simple, unilinear time in which the development of the social formation unfolds obliges Lipietz (1977, pages 21–22) to say that there is no simple, unidimensional space either, and that these correspondences have their own determinate 'topologies'. Thus, for example, "one can speak of the economic space of the capitalist mode of production ... or of the legal space which is superimposed on it", a phrasing which evidently retains Althusser's sense of 'structural causality' and translates it into a spatial lexicon which Vilar would presumably recognise. The most complete of these typologies (translations) has been proposed by Castells (1977), who collapses the dualism between temporality and spatiality to argue that "from the social point of view ... there is no space but an historically-defined *space–time*, a space constructed, worked, practised by social relations" (page 442). Then, still in accord with Althusser, he can claim that (page 125) "all space is constructed" via "the mediation of social practices", which are themselves constituted through "the action of men determined by their particular location in the structure thus defined", and this immediately allows distinctive space–times to be assigned to each level of the social formation. "Thus there will be an ideological space, an institutional space, a space of production, of exchange, of consumption (reproduction), all transforming one another constantly through the class struggle" (table 1).

This last clause picks up Jensen-Butler's comments on a reconstituted time-geography, and it is this (rather than the typology as such) which is important here. But the two are bound together, because Althusser's 'history' is energised through a functionalist teleology, which reduces the specifications of social practice to effects of the structure of the social formation whose own realisation is 'given' by a prior logic of transformation. This has definite consequences for the way in which "the production of space" is understood within this problematic. Let me explain. In *Reading Capital* (Althusser and Balibar, 1970, pages 302–308), transformations are supposed to be brought about through a noncorrespondence (or contradiction) between the forces and relations of production. During phases of transition, the connection between the two "no longer takes the form of a reciprocal limitation, but becomes the transformation of the one by the effect of the other", so that historically "the capitalist nature of the relations of production ... determines and governs the transition of the productive forces to their specifically capitalist form." Thus this primary contradiction has the form (relations of production) → (forces of production), and it is located within the economic level. However, it also reappears as a contradiction between the different levels which, as Giddens (1979a, page 56) puts it, "'reverberate' upon one another to multiply contradictions, which can then 'play back' through the forces/relations contradiction." This is what Althusser means when he describes contradictions within the social formation as being typically 'overdetermined'. Yet they are also essentially *functional* because, as Cutler et al (1977, page 200) have shown, during phases of transition "the mode of intervention of political practice, instead of conserving the limits [of the structure of production] and producing its effects within their determinations, displaces them and transforms them." Clearly, this involves a covert teleology through which men and women become the bearers (*Träger*) of structural determinations whose vectors surface soundlessly in response to sirens which—however

Table 1. Social practices and space-times in the social formation (after Castells, 1977).

Level	Social practice	Space-time	domination	determination
Ideological	legitimation communication	symbolic		
Political/juridicial	integration repression domination regulation	institutional		
Economic	consumption exchange production	consumption transfer production		

distant and distorted—rest on a solidly economic bedrock. For "if the period of transition is brought to an end through the transformation of the productive forces by the relations of production, then the class struggle as such, or the conflict of political forces, can have no independent effectivity. At most, the class struggle performs the role assigned to it by the structure of production"—which is precisely why Glucksmann dismisses Althusser's project as a 'ventriloquist structuralism'. In effect, the economy acts to secure its own conditions of existence, and the scheme collapses into the "closed and empty circle" of functionalism, in which "each component part of the structure exists as an effect of the structure, and it exists because of the function it performs for the structure".

All of this is faithfully mimicked by Castells (1977, page 125). Class struggle becomes the locus of transformation through which "action reacts on the structure itself. It is not simply a vehicle of structured effects: it produces new effects." But Castells insists that these transformations "proceed not from the consciousness of men, but from the specificity of the combinations of their practices ... [which] is determined by the state of the structure." This makes Eliot Hurst's (1980) denunciation of a spatial formalism, which treats spatial structure as the product of an "auto-dominant auto-genesis", plainly self-defeating. It is made from within the supposedly privileged discursive confines of an Althusserian problematic, which contains the identical flaw, projected into the social formation—as though two mirrors were endlessly reflecting one against the other. Further, these images impose themselves on concepts of spatial structure, for Castells (page 442) allows the production of the "new effects" to be represented as a "spatial determination of the social" once this is understood as "a certain efficacity of the social activity expressed in a certain spatial form". This must mean that "spatial forms" are not somehow epiphenomenal, but that they have a definite, and clearly limiting, effectivity. It is through this series of surreptitious equations that *spatial structure becomes both an expression and a means of the realisation of the teleology of structural transformation*. Lipietz (1977) makes this even clearer. In his text (page 22) spatial structure "presents itself at the same time as an articulation of the spaces analysed, as a product, a reflection of the articulation of the social relations, and also ... as an objective constraint which imposes itself on the redeployment [that is, the continuous unfolding] of those social relations."

These are, of course, highly specific constructions, but instances like them can be multiplied many times over. Harvey's (1973) discussion of Piaget's genetic structuralism, for example, prompts him to suggest (page 310) that "it is probable that our culture ... emanates from created space more than it succeeds in creating space ... Neither the activity of space creation nor the final product of created space appear to be within our individual or collective control, but fashioned by forces alien to us."

They are the necessary consequence of dissecting spatial structure through the optic of structuralism, and for all their differences the cumulative weight of these various propositions presses the argument in a single, unmistakable direction. It should now be clear why Olsson talked about the construction of spatial prisons which constrain human actions and social practices. So many of us have been (unwittingly) interned in these theoretical exercise yards—even acting as our own gaolers on occasion— that it is extraordinarily difficult to sum up fairly. But perhaps Braudel's (1966) sentence is exemplary, since it was this great *Annaliste* that Vilar had in mind when he opened his dialogue with Althusser. "When I think of the individual", he wrote at the end of *La Méditerranée*, "I am always inclined to see him imprisoned within a destiny in which he himself has little hand, fixed in a landscape in which the infinite perspectives of the long-term stretch into the distance both behind him and before."

But his is not a solitary confinement. Time-geography refuses the full force of the structuralist critique precisely because it wishes to deflect this antihumanist impetus. Hägerstrand (1978, page 144) has made it perfectly clear that he does "not want to give the impression that ... the pattern of options is absolutely rigid and not responding to the actions of people", so that his notations are intended to carry some sort of emancipatory inflection. He once wrote that the human and social sciences "have as their central problem the conditions of freedom" (Hägerstrand, 1974), and he would therefore presumably want to argue that these will only be disclosed through a reformulation which can take account both of synchorisation and synchronisation—the inescapable necessity for space–time 'packing' in the conduct of practical life. The emphasis on everyday practical life is in fact vital, because Hägerstrand (1970) wants to use time-geography to record "the impact on the ordinary day of the ordinary person". But these critical intentions are again compromised by an insistent functionalism, and the 'ordinary person' readily becomes a polite fiction for the reduction of social practices to the necessary realisation of an inner logic of their constitutive social structures. 'People' in this 'regional science' all too easily become *Träger* for the paths and prisms which contain their trajectories. Thus the distinctions which Carlstein draws from Godelier are inculpated in the same failings that vitiated Althusser's scheme, because, as Giddens has shown, "the teleology of 'contradiction between structure' is that of functional need: the need of the structure or system, unacknowledged by the social actors themselves" (1979a, page 139).

Unfortunately, it is no solution to seize the designative notion of 'project' and wield it like some kind of talisman, a conceptual club to break the circle and rescue the conscious and creative human subject. It clearly has a fundamental role in time-geography, because Hägerstrand (1973, pages 78–79) argues that it is possible to sort out the "chaos of paths and traces" in space and time by ordering them into coherent, hierarchically arrayed clusters—that is, projects—which represent the steps necessary for movement towards determinate goals. Then "each project would seem to try, between its beginning and its end, to accommodate its parts, be they tangible or not, in the surrounding maze of free paths and open space–times left over by other projects or gained through competition with them", so that *the inter-locking of projects of different life-spans, up and down the hierarchy and between hierarchies, is the central problem for analysis*." But when set out like this projects take on a life of their own, and social relations are reduced to a form of what Sartre (1976) called 'seriality' (Aronson, 1980; Poster, 1979). The parallel bites even deeper than this, for Sartre connects seriality to *scarcity*, and this is also a *Leitmotiv* of time-geography. Nevertheless, Carlstein believes that "by thus incorporating human intentionality and normative action—the trajectories of mind and matter inherent in the concept of project—we ensure that ... human individuals are neither relegated to be the staged puppets in a reductionist drama, nor the forlorn subjects to an endless array of negative constraints" (1980, page 249).

But intentionality and normative action are *not* synonyms, and the relations between them need to be elucidated rather than elided. Projects are not the "empirically-given, pre-existing structures" which van Paassen (1976) detects in Hägerstrand's writings; but neither can they be subsumed within a system of positive 'steering constraints' and thereby made *immediately* conformable to an underlying structure of legitimation. What we need to know, as Olsson might say, is how this "change of sign" —from negative to positive—is effected and sustained. And it is partly for these reasons that both Buttimer (1976) and Mårtensson (1979) have called for an exploration of the connections between classical time-geography and constitutive phenomenology, because Hägerstrand's scheme says little about the *constitution* of projects—as I have indicated elsewhere (1978b), what matters most is their sequential order—whereas Schutz leaves scant space for their *completion*. Yet the bonds between the two must be retained, and even strengthened, if we are to avoid what Giddens calls a conceptual 'cutting into' the continuity of action. But they are occluded in most of these writings, and as a result much of time-geography has been preoccupied with *behaviour* rather than with *action*, and in its uncertain oscillation between physicalism and idealism it has largely failed to describe the recursive engagements of interaction and structure. Certainly, when Hägerstrand (1970, page 20) talks about a "flow of life-paths controlled by *given* capabilities and moving through a system of *outside* constraints" (my italics), then, to speak with Steiner (1975, page 217), he condemns the social actors in his series to "turn forever on the treadmill of the present". For how can their actions then have any effectivity other than to ensure the continuous, routinised reproduction of an "enormous maze ... about which [they] can do very little"? (Hägerstrand, 1970, page 18).

This spatial metaphor is a deliberate one, of course, and my appeal to Steiner is not casual either, since the linguistic categories of time-geography themselves mark out a functionalist enclosure. The very concept of 'space' which this contains resonates with what Hägerstrand calls "an inward-directed finitude" (otherwise, of course, it could hardly carry the constellation of meanings required of its definition as a scarce resource on which individuals and groups have to draw in order to complete their projects). In a thoughtful commentary on the words *space* in English and *rum* in Swedish, Gould (1981) has confirmed that in Hägerstrand's usage "the emphasis of meaning is upon constraint, packing, allowing and forbidding", and that its domain is "the adult world where my actions constrain yours, where our interdependencies jostle and compete, demand and require." In this *rum*-time world, he concludes, "we have lost the innocent freedom of the child in the sun-lit meadow", capering through the infinitude, openness, boundlessness connoted by the English sense of *space*. In time-geography too, spatial structure finally seems to close in and close off.

And so we are returned to the prison-house of language. But this is not inappropriate, for the invariant message of all these heterogeneous codes is that the very notion of the human subject is problematic, and that this is in some way inscribed in discourse itself. Perhaps the most challenging presentation of these themes is still that contained in the poststructuralist writings of Foucault, whose archaeology of the human sciences proposed "man" as "an invention of recent date", which emerged through a series of transformations in the discursive space opened up by Kant's *Critique of Pure Reason* and which is now "perhaps nearing its end" (Foucault, 1970, page 387; Sheridan, 1980). Lemert (1979, page 229) also has appealed to these theses, together with those of Derrida, to foreclose what he calls "homocentrism"—that is, "that discursive formation which centres itself upon man as a finite [subject] who dominates his own history"—and to demonstrate why "it is no surprise that a [social science] living and working in the twilight of man has begun cautiously to turn to language." Although there is no space (*rum*?) to explicate those arguments here, we can use them to suggest that the deficiencies we have noted in the various formulations spelled out in this section will not, and cannot, be repaired through the reinstatement of an exuberant voluntarism, which promotes agency over structure, and treats spatial formation as the direct unfolding of "purposeful social practice" (Soja, 1980, page 210)[1]. To repeat: the 'linguistic turn' has been a *necessary* manoeuvre, and it is of the first importance to retain some, but only some, elements of the structuralism which has spiraled through it.

[1] Soja's (1980) important essay is a considered attempt to reinstate the perspectives opened out by Lefebvre's humanist Marxism, which is itself a rejection of the structural Marxism of Althusser.

Only then will it be possible to acknowledge—to concede if not to consent—the constitutive function of spatial structure in the *société disciplinaire* (Foucault, 1970; Gregory, 1980).

Emancipation and the spatial structure of social relations

I now want to begin to weave these strands together, and to suggest that the explication and contestation of the *société disciplinaire*—the unfolding of the emancipatory interest—depends upon connecting these discussions to a philosophy of transcendental realism, whose central concept of *ontological depth* allows for the recognition of the multitiered stratification of reality in terms which are congruent with the notion of 'interactive structure'. This case has been argued in detail by Bhaskar (1978), who claims that the human sciences become "intrinsically critical and self-critical" once they are reformulated to accommodate what he calls a *duality of praxis* and a *duality of structure* (Bhaskar, 1979; 1980). I will decode this couplet by conjoining Bhaskar's account to the theory of structuration proposed by Giddens (1979a), with which it has important affinities, and then indicate some deficiencies in Giddens's programmatic reconstruction of social theory.

The connections between these texts, and between them and Habermas's project, form a discursive 'triangle'. I will not provide a rigorous exegesis of these writings here, and I definitely do *not* want to imply that their vertices can be mapped directly onto each other either through rotation or translation. Nor do I wish to endorse them in their entirety, but rather use the intersections between them to develop a series of specifications of spatial structure which, I shall claim, necessitates a further reworking of the intersections between time-geography and structural Marxism. I shall not seek to provide that here, and my comments will inevitably be highly abstract at this stage because, like Habermas (1979), I am concerned here with the "logic" of "structure-forming processes" and not with the development of their "empirical substrates". However, this does *not* mean that these are in any way unimportant and that I regard them either as parasitic or as residual. Although the sketch which follows is coloured by the structuralist hues which have been drawn out of the preceding materials, it is certainly *not* the case that, as Johnston (1980) has suggested, "the detailed morphology of [regional] mosaics" is obscured by this frame of reference. On the contrary, it is the incorporation of ontological depth through a realist philosophy of science which makes possible the excavation of successive 'layers' of social practice and the disclosure of the spatial structure of their constitutive social relations. This is something which positivism—if it is defined correctly—simply *cannot* provide, and any appeals to a presumed 'didactic relevance' fail to save it from incoherence.

Giddens (1976, page 153) is not indifferent to postpositivist philosophy, of course, and his theorems can be shown to flow from a series of engagements with the contemporary critique of positivism. But in the course of this he assimilates positivism to *naturalism*, "the thesis that the logical frameworks of natural and social science are in essential respects the same", and attributes the enduring 'pathos' of nineteenth-century European social thought to their historical conjuncture. This means that he recognises what Frisby calls the "affirmative impulse of critical enlightenment" which informed the traditional (Comtean) model, namely that "the extension of natural science to the study of man was undertaken with the promise of liberating human beings from their bondage to forces perceived only dimly or in mystified form." But it also obliges him to claim that, within such a tradition—whose residues have proved to be remarkably resilient—the "reciprocal relation between social analysis and everyday conduct is represented only in marginal forms", because its prescriptions derive from the discovery of "social laws" which, although they are nonrefractory, are nevertheless supposed to be invariant and immutable. The space for purposeful social practice is thus sharply circumscribed, and in seeking to rescue the conscious and creative subject, to widen the circle of social effectivity, Giddens is prompted to reject positivism (which is surely right) *and therefore naturalism* (which is much more problematic). It is important to state, therefore, that these retrospective identities need to be prised apart[2]. As Giddens has subsequently admitted, this "received model of the natural sciences" is *not* an appropriate formulation of nomological explanation either in the natural or in the social sciences (the same objection can be registered against Habermas's representation of "empirical-analytic science"), and these are *not* "two independently constituted forms of intellectual endeavour" but instead "feed from a pool of common problems". Indeed, Giddens (1977) now accepts that their solution will require "a detailed reworking of pre-existing formulations of realism", which would in turn almost certainly require a reworking of his theory of structuration —although he has *not* provided one. Nevertheless, it would be consistent to do so because, as I have argued elsewhere (Gregory, 1978b, pages 56– 59), a realist philosophy of science is, in principle, capable of securing the 'network model' of scientific inference which Giddens has explicitly endorsed. In fact, it is rather more urgent than this, because such a postempiricist reformulation both acknowledges and accommodates theoretical and empirical 'co-determination', and its methodological protocols can therefore be used to connect scientific practice to 'natural language' and hence to a properly emancipatory political practice.

This is not to say that realism constitutes a preformed philosophy for the human and social sciences, of course, but it is to suggest, with Bhaskar, that realism permits them to be sciences in *exactly the same sense* though not in *exactly the same way* as the natural sciences. This matters not for

[2] Although their conflation was a mistake I also made myself (1978b).

any privileges which might then be accorded to their discursive claims, but rather for the possibility of critique—"the passage from explanatory theories to practical imperatives"—which it opens up. These claims are necessarily limited by what Bhaskar (1980) has identified as the activity dependence, concept dependence, and space–time dependence of social forms, but they can nevertheless be shown to derive from the specification of 'emergent laws'. It then follows that realism has to be distanced from an essentialism in which societies are represented as expressive totalities, and whose social practices are given by, or reducible to, an inescapable 'inner logic' of their constitutive social formations.

It is in order to avoid these essentialist errors that Urry (1981) has tried to provide a set of "concepts which designate the social space in which individual subjectivities are constituted and reproduced", and which remain broadly consonant with the precepts of historical materialism. It is self-evidently true that Marx, especially in his later writings, was primarily concerned with the social relations of production through which individuals 'bear' objective functions—that is, with the *Träger* which reappear in structural Marxism. But Urry insists that this does not proscribe, *but on the contrary presupposes*, a conjoint realm of social practices through which individuals are intersubjectively constituted and reproduced, and which are characteristically the concern of a humanistic Marxism. He therefore develops a categorical distinction between "the social relations between agents in the capitalist economy, and social interactions between subjects within civil society". He argues (Urry, 1981, page 72):

"The process of constitution of subjectivity is achieved through language, through the positioning it accords to individuals within particular discursive formations. It is not merely that individuals are allocated roles, since this implies that they are already formed as subjects. It is that through social experience, through their position within various discourses, individuals view themselves and act as autonomous centres of creativity, consciousness and initiative. They are thus determined as conscious and self-reflexive, to act as autonomous, whole and independent subjects. The most important interpellations of the subject are those of spatio–temporal location and of gender. The effect of the former is that individual subjects are constituted who are aware of their presence as subjects resident within a particular spatial location (street, town/countryside, region, nation) at a given period of time (born of a particular generation defined by its place in relation to others). The interpellation of gender is to produce autonomous sex-ed subjects, each defined by its relationship of difference with the other. Gender, like spatio–temporal location, is socially constructed and produced through the differences entailed with language."

The importance of spatio–temporal interpellation [which corresponds roughly to the connections between Harvey's (1973, page 23) 'geographical' and 'sociological' imaginations] has, of course, been registered by humanistic geography, although in a different vocabulary, and Tuan (1977) and others have made some progress in opening up the affective passages between 'space' and 'place'. But the difference is more than merely lexical, because these social constructions have a social effectivity and, as Baker (1981) reminds us, "attachment to—and alienation from—place or locale is an integral part of the process of social structuration."

Although his trajectory is different—and I will return to this shortly—Giddens intends his theory of structuration "to promote a recovery of the subject without lapsing into subjectivism" and to grasp the space–time intersections "inherent in the constitution of all social interaction". His scheme is summarised in figure 1. Here, intersubjective relations are organised as *systems of interaction* through a series of regular, recurrent, and reproduced social practices, all of which disclose three fundamental dimensions: the communication of meaning, the exercise of power, and the evaluative judgement of conduct. These are constituted by actors routinely drawing upon the generative rules and resources—that is, the *modalities* of interpretative schemes, facilities, and norms—made available by *structures* of signification, domination, and legitimation, in such a way that the successive and simultaneous engagement of these modalities necessarily reconstitutes their constitutive structures. Social life thus displays a "recursiveness" in which "the structural properties of social systems are both the medium and the outcome of the practices that constitute those systems" (Giddens, 1979a, pages 69–94). An example should make this clear. When I speak, I necessarily draw upon a preexisting linguistic structure, and although I might not be able to specify the rules and resources which this makes available with any precision (particularly at levels below elementary grammar and syntax), its existence is nevertheless a (typically unacknowledged) *condition* of every intelligible speech act. Symmetrically, these utterances necessarily reach back to reconstitute that structure, whose reproduction thus becomes an unintended *consequence* of every speech act.

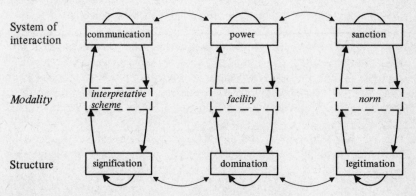

Figure 1. A model of structuration (after Giddens, 1979a).

These phrasings evidently treat structure not as a barrier to, or a constraint upon, action, but instead as essentially involved in its production. As Bhaskar (1980, page 18) puts it, "society is both ever-present condition and continually reproduced outcome of human agency: this is the duality of structure." To admit the existence of *unacknowledged* conditions and *unintended* consequences in this way demands a move towards a form of structural explanation, and Giddens represents structure as "an absent set of differences, temporally 'present' only in their instantiation, in the constituting moments of social systems" (1979a, page 64). But he has also on occasion (1976, page 127) claimed that "structures only exist as the reproduced conduct of situated actors with definite intentions and interests", which allowed Layder (1981) to object, quite properly, that such an identification of structure "with the doings or productions of interactants (instantiated rules and resources)" necessarily "emasculates" the concept "by depriving it of autonomous properties or a pre-given facility." The effect of such an absence, he argued, is thus "to drive the notion of structure back into the given, the concrete"[3]. This objection is an important one, even though Giddens soon corrected the lapse to say that in his view structures *cannot* be treated "as the situated doings of concrete subjects, which they both serve to constitute and are constituted by" (1976, page 128), because they are characterised by the absence of a subject and exist (as it were) paradigmatically "out of space and time" as "relations of presence and absence recursively ordered" (1979a, page 255). I take this to mean that these structures have to be reconstructed through a collation of their instantiations: that is, they are not immediately and extensively 'present' in any one of them as Layder seems to think. Structures are not 'given' in this or that particular, therefore, but their engagements are differentiated in space and time—which is to say that they are *distanciated*—so that conjunctural sequences of instantiation are both bounded *and in some sense contingent*. Certainly Giddens repeatedly insists that action is not merely the mechanical outcome of generative structures, and that, in drawing upon the rules and resources which they make available, actors are necessarily displaying varying degrees of 'penetration' of practical life. These variant knowledges are not incidental to the conduct of everyday life, because actions are intrinsically reflexive —they are motivated, monitored, and rationalised—and as such are essentially involved in the reproduction of social structures. In Bhaskar's terms, "human agency is thus both production and reproduction of the conditions of production, including society: this is the duality of praxis" (1980, page 18).

To say this is definitely not to endorse the 'individualist' and 'voluntarist' inflections which some critics have read into the scheme.

[3] Layder's (1981) reconstructions are, I think, based upon a misreading of Giddens's theorems (although this is easy enough!), and a number of objections can be made to them, because they say even less about concepts of determination.

It might be possible to show that, even though his early formalisations emerged through a *critique* of interpretative sociologies, Giddens was nevertheless either unable or unwilling to acknowledge the full extent of the objections which could be registered against subjectivism[4]. Thus Clegg (1979) has claimed to find in those writings "an Hegelian emphasis on action" rather than, as he would undoubtedly prefer, a "Marxian" emphasis on "the properties of action as themselves an 'effect of the structures in the field of social relations' "—although the unqualified attribution of these positions to Hegel and Marx is itself open to question. It may well be true that in some respects Giddens's scheme fails "to free structures from the fiction of the subject", but it is far from clear that the substitutions which Clegg makes (pages 68–75, cf.98–99) (in part through a series of appropriations from Foucault), succeed in freeing action from the equally fictive teleology of the system. It is extremely important to make this *double* movement, although Foucault says much more about the former than the latter. The interpenetrations of human agency and social structure are in fact so unstable that any attempt to mirror the movement between them is bound to be a precarious struggle which can collapse at any moment. Indeed, Dawe (1978) has claimed that the entire history of social theory, whatever its ostensible intentions, can be written as a variation on "the recurring theme of the negation of human agency". To the extent that this is even true of the heterogeneous 'sociologies of action', as Dawe maintains, it becomes doubly necessary to avoid the substitution of one ontology for another. This has to be said because S Williams (1981) has argued, in an otherwise wholly admirable essay, that a realist human geography will require the replacement of an "ontology of events" by an "ontology of structures", whereas I prefer to follow Layder in identifying "the relations between different ontological domains at the same time as recognising their integrity as differentiated features of social reality" (1981, page 141). To repeat: it is a mistake, a one-dimensional reduction, to regard either actions or structures as more 'real' than the other and, like Giddens, I consider it a task of the utmost importance to overcome this deeply sedimented dualism.

It follows from this that the 'transformational model' of society—Bhaskar's description—logically entails a concept of *articulation*, and I now want to show that this demands a double concept of spatial structure. Its *first* specification is in the *immediate* sense of the 'binding of space and time' in the modalities which connect systems to structures. Although Giddens (1979a, page 205) drew upon time-geography to underscore "the co-ordination of movement in time and space in social activity, as the coupling of a multiplicity of paths or trajectories", Pred (1981) has argued that his remarks are inadequate because they fail to explicate "the means by which the everyday shaping and reproduction of self and society, of

[4] His *New Rules of Sociological Method* (1976) is especially vulnerable in this respect.

individual and institution, come to be expressed as *specific* structure-influenced and structure-influencing *practices occurring at determinate locations in time and space*." In order to fill this gap, Pred proposes as a central theorem that "without the constant channelling of individual paths into and out of the activity bundles of organizational and institutional projects, there are no 'social practices', society has no observable workings, and society is without a generating and perpetuating structure". But although these recurrent space–time intersections, which Pred takes directly from Hägerstrand (Giddens was more cautious), constitute a necessary medium of social reproduction, they do not provide a sufficient condition for the constitution of social structure. It is important to emphasise this because on Pred's reading, if "social reproduction and the dialectics of practice and structure can only be expressed through an unending sequence of temporally and spatially specific intersections between individual paths and institutional projects", then it follows that "social transformations and altered structural relations ['institutional interrelations'] only occur through the introduction, disappearance or modification of institutional projects. Or, social change and altered structural relations can only occur through the addition, elimination or recasting of the temporally and spatially specific path couplings demanded by institutions from day to day" (page 17). I wish to claim, in the strongest terms, that this is *not* the case: that although space–time intersections may be a means through which structural transformations are effected, space–time webs are neither completely identified with, nor are they fully determined by, specific structural relations. And I agree with Layder that to suggest otherwise is to drive structure back into the concrete of everyday life, and to inscribe its moments directly in the surficial geometry of spatial interaction. To be sure, Giddens acknowledges the importance of 'routine' in ensuring the continuity of social reproduction —and this displays definite space–time rhythms—so that social change can be explicated through an identification of "the conditions under which the routinised character of social interaction is sustained or dislocated" (1979a, page 219). But these conditions are *not* exhausted by the delineation of space–time webs. Any attempt to make them so threatens, *inter alia*, to make the concept of spatial structure equivalent to, and delimited by, a system of spatial interaction. If we are to avoid such a premature closure, we shall have to specify (1) a hierarchy of structural domains, and (2) their structural conditions of existence. These twin imperatives are in some degree latent both within time-geography and the theory of structuration, of course, but they are elided in Pred's (1981) subsequent discussion of 'dominant projects'. In contrast, their explication demands conjoining this first concept of spatial structure to a second.

This *second* specification is in the *distanciated* sense of "the uneven development of different sectors or regions of social systems". I wish to propose that the contingent and differential engagement of the modalities of structure *depends upon* and is *structured by* a definite spatial structure of social relations. This connects back to Giddens's (1979a) discussions of the presence/absence couple, and especially to his view that "the differences which constitute social systems represent a dialectic of presences and absences in space and time" which "are only brought into being and reproduced via the virtual order of differences of structures, expressed in the duality of structure" (page 71). But Bhaskar provides a more accessible and acceptable *entrée* to these notions by arguing that the transformational model further requires "a system of mediating concepts, encompassing both aspects of the duality of praxis, designating the 'slots', as it were, in the social structure into which active subjects must slip in order to reproduce it; that is, a system of concepts designating the 'point of contact' between human agency and social structures. Such a point, linking action to structure, must both endure and be immediately occupied by individuals." The point-patterns which these articulations trace out are defined by a *position–practice system* which, Bhaskar (1979) insists, "must be conceptualised as holding between the positions and practices (or better, positioned-practices), not between the individuals who occupy/ engage in them." In other words—and this is the crux of the matter—it has to be located *not at the level of social interactions but at the level of social relations*. This distinction, which evidently parallels Urry's, immediately opens up "the range of questions having to do with the *distribution* of the conditions of action" (page 52), questions which are central to the second specification of spatial structure (and which is, in part, why I still think we have something to learn from a reexamination of the texts of the locational school). While the answers to these questions are clearly not independent of the space–time webs of classical time-geography, neither are they synonymous with them. That they are, nevertheless, bound in to what I previously called 'spatial formation', and that as such they serve to reactivate the problematic of structural Marxism, can be demonstrated by returning to Giddens's original formulation and, in effect, stiffening its structural inflections. For Giddens also acknowledges that it is *practices* "which (via the duality of structure) have to be regarded as the 'points of articulation' between actors and structures", and he sees "no difficulty" in thinking of "social systems as structured 'fields' in which (as reproduced in the temporality of interaction) actors occupy definite *positions* vis-à-vis one another" and which carry definite "prerogatives and obligations" for their "incumbents". The theoretical status of this system is far from easy to establish, and it needs much more elaboration, but insofar as Giddens situates social practices within "intersecting sets of rules and resources"—so that they skewer all three modalities—then they must derive from the structural domains identified in table 2 (Giddens, 1979a, pages 82, 117). It is these which need to be strengthened, because they constitute the matrices within which social relations are to be theorised. And in fact this typology has such close affinities with the classification of social practices and space-times developed from structural Marxism (table 1) (although I think it avoids Althusser's residual functionalism), that it can usefully be used to map out the conceptual junction between class relations and spatial structure.

For Giddens, "class refers above all to a mode of institutional organisation", which means that it is necessary "to connect, via the duality of structure, a theory of class society, as an institutional order, with an account of how class relations are expressed in concrete types of group formation and consciousness" and instantiated in systems of social interaction (1979a, page 110). In his early writings, Giddens (1979b)[5] distinguished between 'mediate' and 'proximate' *sources* of structuration— generalised and localised bases for the emergence of structured class formations—and then simply superimposed this grid over the recursive *sequence* of structuration (pages 105-112). In this way, he was able to represent "class structure [as] both the medium and the outcome of social reproduction" and to say that its three constitutive dimensions of legitimation, domination, and signification are "routinely drawn upon by actors in the course of constituting class relations as interactions; in drawing upon them as modalities of interaction, they also reproduce them as that structure" (Giddens, 1976, page 123). Some commentators have alleged that this presentation is grounded in a series of *market encounters*, and that it accords a substantive priority to *human agency*: in short, that it is neo-Weberian (Binns, 1977; Crompton and Gubbay, 1977). Giddens (1979b, page 297) dismisses this as an "almost gratuitous misreading" of his position; intellectual fideism is not the issue here, and I think Giddens is right to say that he is in fact much closer to Marx than his critics allow. But I also think that his construction *is* deficient when stated like this, because it contains no concept of *determination*, and this is of decisive importance.

His later writings have clarified this considerably, however, provided that they are placed in the context of his argument as a whole—and he

Table 2. Theoretical components of structuration (after Giddens, 1979a, figures 3.1, 3.2).

System : structure	Theoretical system	Domain
communication : signification	theory of coding	symbolic orders/modes of discourse
power : domination	theory of resource authorisation	political institutions
	theory of resource allocation	economic institutions
sanction : legitimation	theory of normative regulation	legal orders/modes of sanction

[5] This text was originally published in 1973, but all my references are to the 1979 edition which contains an important postscript.

does not make this easy. Nevertheless, it is possible to show that these have begun to provide for the specification of (1) a hierarchy of structural domains and (2) their structural conditions of existence. This is achieved through a critical appropriation of historical materialism, and in particular by a theorisation of the society–nature relation. Here too Giddens is not so very far from Habermas, whose own reconstruction of historical materialism attempts to explicate the 'organisation principles' of different social formations—abstract regulations which "circumscribe ranges of possibility" but which do not uniquely determine particular combinations of modes of production—by relating the 'institutional cores' around which different relations of production crystallise to 'new categories of burden' ('problem situations') which they successively and determinately entrain. These form "a spectrum of problems connected with the self-constitution of society, ranging from demarcation in relation to the environment, through self-regulation and self-regulated exchange with external nature, to self-regulated exchange with internal nature. With each evolutionarily new problem situation there arise new scarcities: scarcities of technically feasible power, politically established security, economically produced value and culturally supplied meaning" (Habermas, 1979, pages 164–167). The logic of this scheme need not detain us, although its structuralist filiations are remarkable. What is more important here is the *complexity* of terrain which is spelled out and, in consequence, the need for discussions to be sufficiently nuanced to distinguish a *nexus* of society–nature interpenetrations. Whether this can be unfolded into an evolutionary progression is an open question, but its opacity to the linguistic categories of environmental determinism and possibilism is clear enough. These have, historically, supported what R Williams (1980, page 107) calls a theoretical 'deformation': for, although the "far and middle reaches of our material environment" may be beyond human election or conscious control, "what can then properly be described as an 'external situation' modulates, in complex ways, into what is already an 'interactive situation' and then, crucially, into an area of material conditions in which it is wholly unreasonable to speak of 'nature' as distinct from 'man'". In other words, an adequate theory of structuration must grasp the constituted *materiality* of practical life in all its foliations (Sayer, 1979; Smith, 1980).

Giddens (1979a) therefore proposes that all societies exist in a contradictory relation to nature, but one which "is always mediated by the institutions in terms of which, in the duality of structure, social reproduction is carried on. The existential contradiction of human existence thus becomes translated into structural contradiction, which is really its only medium" (page 141). Structural contradiction is then defined as an opposition or disjunction of structural principles of social systems which operate in terms of each other *and* in contravention to each other; whereas primary contradictions—which connect up to the 'conditions of existence' I referred to earlier—are those which are "fundamentally and inextricably" involved in social reproduction, "not on a functional basis, but because they enter into the very structuring of what that system *is*" (page 161). A number of writers have, of course, made much the same point, but I do not want to follow those who have directly identified a series of 'spatial contradictions' (which typically cluster around notions of combined and uneven development), because in the form in which they are usually stated they resuscitate the Kantian dualism which separates space as an autonomous realm of existence (Soja and Hadjimichalis, 1979; Peet, 1978)[6]. Neither do I want to say that "the theory of nature holds within it an integral theory of space", because I do not think that the one unfolds synecdochically out of the other (Smith and O'Keefe, 1980). I simply want to suggest, with Dunford (1980, page 73), that "the concept of nature as a complex of natural and social *conditions* of production and as a product [or *consequence*] of human intervention and human activity plays a central role in understanding the structure of geographical space." As such, the labour process (properly conceived) is the very locus of structuration since, in Marx's words, "labour is, first of all, a process between man and nature" through which "he acts upon external nature and changes it, and in this way he simultaneously changes his own nature" (Marx, 1976, page 283; Gregory, 1981b). This double movement or 'contradictory unity', with all its precursive evocations of the Vidalian school, is seized upon by Giddens to draw out the richly-textured web of mediations which it necessarily entails (and which received such short shrift in *la géographie humaine*). But a hierarchy of effects and determinations remains to be established, and the difficulties which are embedded in this complex question can be clarified through a more detailed specification of the hierarchy of structural domains and their structural conditions of existence which typify capitalism. More particularly, therefore, Giddens argues that in capitalist societies:

(1) *resource allocation* (the economic dimension of domination) has a definite primacy, because capitalism "turns the exploitation of nature into a propelling force of social change". Its development "sets under way an impetus to continuous technical innovation and the expansion of the productive forces: this is 'autonomous' in the sense that the expanded reproduction of capitalism is promoted by the very operation of capitalist production itself." However, for a variety of reasons this is chronically unstable, so:

(2) a primary contradiction occurs between *private appropriation and socialised production*. While this has taken a number of different and historically-specific forms, this is in itself an indication of the internally convulsive composition of the capitalist mode of production (which is further accentuated through a series of secondary contradictions).

---

[6] But for a critical commentary see Smith (1981).

Both of these propositions clearly require very considerable elaboration (and, I think, emendation: the logic of capitalist accumulation is far from simple), but taken together they imply the following (Giddens, 1979a, pages 163–164):

"In capitalist enterprise property becomes both the organising principle of production at the same time as it is the source of class division. Only in capitalism are the sources of contradiction and class conflict identical. Ownership of private property is both the means of appropriating a surplus product ... and simultaneously the means whereby the economic system is mobilised. This is why Marx's stress upon the process whereby labour power becomes a commodity is so important; for it is in the labour contract that contradiction and class conflict, in the capitalist mode of production, coincide."

This clearly distances Giddens from Weber; but there is still some way to go. For although these theorems, especially in their final moments, closely resemble the Althusserian account of the 'real' subordination of labour to capital[7], their location within the problematic of structuration means that it *must* be possible to explicate the ways in which specific structures of signification and legitimation are also 'mobilised' in order to confirm (or contest) the reproduction of specific structures of domination, *without collapsing into an economism*. Mobilisation, effected through the changing modalities of structuration, is always contingent, and its final consequences can be fully determined *neither by the economy nor by the self-determination of economic agents*.

What I wish to emphasise, therefore, is that the economy has definite political, cultural, and legal conditions of existence, but that it cannot secure them (as it were) 'autogenetically'. Giddens himself suggests that all three aspects of structure—signification, domination, and legitimation— can be understood in terms of the "mediations and transformations which they make possible in the temporal–spatial constitution of social systems", and that these concern what he calls the *convertibility relations* of rules *and resources*, such as the cascade: private property → money → capital → labour contract → profit (1979a, page 103). There are some obvious difficulties in the theoretical formation and articulation of these chains, but however, these are to be formulated they involve (but do not uniquely determine) the space–time intersections delineated by time-geography, and they depend, *crucially*, on the distribution of the conditions of production —and hence on the spatial structure of social relations. If we follow Hindess and Hirst (1977, page 65) to define means of production as "all the conditions necessary to the operation of a particular labour process which are combined in the units of production in which that process takes place", then we can say that "if any of these conditions is exclusively possessed by a definite category of agents, and the agents who direct or

---

[7] I have discussed this in detail in Gregory (1982a).

operate the labour process are separated from them, then such relations [that is, of 'presence' and 'absence', 'possession' and 'separation'] provide the basis for class relations." What this means, in turn, is that *relations between the 'unit of production' or 'enterprise' and the systems of circulation or distribution of the conditions of production must be analysed if class relations are to be rigorously determined*" (Cutler et al, 1977, page 252). This clearly underwrites the importance of revitalising location theory, and to this extent, *and in this sense*, it ought to be a matter of concern that even sympathetic reviewers like Wood (1978) now "suspend judgement" over whether locational analysis will "retain the central role in human geography that seemed so natural a few years ago." But it also means that its intellectual isolation, which its founding fathers themselves frequently rejected[8], must be irredeemably compromised as its concepts are connected up to social theory to provide for the 'resocialisation' of the space-economy. In this regard I think Giddens's claims for the autonomy of the economy are overstated, for access to the means of production can be closed off in registers other than the economic (Gregory, 1982a). To put it somewhat differently, this suggests that the interconnections between the space-times set out in table 1 (and, by implication, in table 2) are of strategic importance.

Of course, there is no reason to suppose that these various grids are fully congruent, and that their rotations completely mesh, but properly understood these interconnections do not pose *functional* questions at all. These scale-specific differentiations are bound into the structuration of the space-economy, and their demarcations are confirmed and contested through a host of different social practices, just as their mediations are effected through a whole series of different convertibility relations. But we need not see the patterning of these practices and relations in time and space as only a planar prison which ensnares and encloses biography and history; rather, the 'looseness of fit' between the different space-times can be a vital means of escape and emancipation. The interstices in the *société disciplinaire*—interstices which splinter the conventional disciplinary matrix—can, perhaps, be opened out to widen the 'curve of contingency'. Thus, we can step outside the formal surveillance of space-time partitions and intersections and interrogate, both in intellectual and practical terms, the spatial structure of social relations which lies hidden beneath them and which is at once their condition and their consequence. When phrased like this, the analysis of spatial structure is not derivative of, or secondary to, the analysis of social structure (as the structuralist critique tacitly assumes), because it is utterly impossible to theorise the one without the other. And this conjoint effort is intrinsically emancipatory, for spatial structure is then not merely the arena within which class conflicts and structural contradictions converge and 'express'

[8] I have provided a detailed discussion in Gregory (1981a), and a more general survey in Gregory (1982b).

themselves, but also the domain—the heterogeneous *collage* of domains—
in which, and in part *through* which, social relations are continually
constituted, reproduced, and transformed.

References

Althusser L, Balibar E, 1970 *Reading Capital* (New Left Books, London)

Aronson R, 1980 *Jean-Paul Sartre: Philosophy in the World* (New Left Books, London)

Baker A, 1981 "An historico-geographical perspective on time and space and on period and place" *Progress in Human Geography* 5 439–443

Bhaskar R, 1978 *A Realist Theory of Science* (Harvester Press, Brighton, Sussex)

Bhaskar R, 1979 *The Possibility of Naturalism.: A Philosophical Critique of the Contemporary Human Sciences* (Harvester Press, Brighton, Sussex)

Bhaskar R, 1980 "Scientific explanation and human emancipation" *Radical Philosophy* **26** 16–28

Binns D, 1977 *Beyond the Sociology of Conflict* (Macmillan, London)

Blaut J, 1961 "Space and process" *The Professional Geographer* **13** 1–7

Braudel F, 1966 *La Mediterranée et le Monde Mediterranéen* (Librairie Armand Colin, Paris)

Bunge W, 1962 *Theoretical Geography* (Gleerup, Lund)

Buttimer A, 1976 "Grasping the dynamism of the life-world" *Annals of the Association of American Geographers* **66** 277–292

Carlstein T, 1980 *Time Resources, Society and Ecology* Department of Geography, Royal University of Lund, Lund

Castells M, 1977 *The Urban Question* (Edward Arnold, London)

Clegg S, 1979 *The Theory of Power and Organization* (Routledge and Kegan Paul, Henley-on-Thames, Oxon)

Cliff A, Ord J K, 1981 *Spatial Processes: Models and Applications* (Pion, London)

Crompton R, Gubbay J, 1977 *Economy and Class Structure* (Macmillan, London)

Cutler A, Hindess B, Hirst P, Hussain A, 1977 *Marx's Capital and Capitalism Today* (Routledge and Kegan Paul, Henley-on-Thames, Oxon)

Dawe A, 1978 "Theories of social action" in *A History of Sociological Analysis* Eds T Bottomore, R Nisbet (Heinemann Educational Books, London) pp 362–417

Dunford M, 1980 "Historical materialism and geography" in Research Papers in Geography 73, Department of Geography, University of Sussex, Brighton

Foucault M, 1970 *The Order of Things: An Archaeology of the Human Sciences* (Tavistock Publications, London)

Friedman J, 1974 "Marxism, structuralism and vulgar materialism" *Man* **9** 444–469

Giddens A, 1976 *New Rules of Sociological Method* (Hutchinson, London)

Giddens A, 1977 *Studies in Social and Political Theory* (Hutchinson, London)

Giddens A, 1979a *Central Problems in Social Theory* (Macmillan, London)

Giddens A, 1979b *The Class Structure of the Advanced Societies* second edition (Hutchinson, London)

Goddard D, 1972 "Anthropology: the limits of functionalism" in *Ideology in Social Science* Ed. R Blackburn (Fontana, London) pp 61–75

Godelier M, 1972 *Rationality and Irrationality in Economics* (New Left Books, London)

Gould P, 1974 "Some Steineresque comments and Monodian asides on geography in Europe" *Geoforum* **17** 9–13

Gould P, 1980 "*Q*-analysis, or a language of structure: an introduction for social scientists, geographers and planners" *International Journal of Man-Machine Studies* **12** 169–199

Gould P, 1981 "Space and rum: an English note on espacien and rumian meaning" *Geografiskar Annaler* **63B** 1–3

Gould P, 1982 "Is it necessary to choose? Some technical, hermeneutic and emancipatory thoughts on inquiry" this volume, pp 71–104

Gregory D, 1978a "The discourse of the past: phenomenology, structuralism and historical geography" *Journal of Historical Geography* **4** 161–173

Gregory D, 1978b *Ideology, Science and Human Geography* (Hutchinson, London)

Gregory D, 1980 "The ideology of control: systems theory and geography" *Tijdschrift voor Economische en Social Geografie* **71** 327–342

Gregory D, 1981a "Alfred Weber and location theory" in *Geography, Ideology and Social Concern* Ed. D Stoddart (Basil Blackwell, Oxford) pp 165–185

Gregory D, 1981b "Human agency and human geography" *Transactions of the Institute of British Geographers* **6** 1–18

Gregory D, 1982a *Regional Transformation and Industrial Revolution: A Geography of the Yorkshire Woollen Industry, 1780–1840* (Macmillan, London)

Gregory D, 1982b *Social Theory and Spatial Structure* (Hutchinson, London)

Habermas J, 1979 *Communication and the Evolution of Society* (Heinemann Educational Books, London)

Hägerstrand T, 1970 "What about people in regional science?" *Papers of the Regional Science Association* **24** 7–21

Hägerstrand T, 1973 "The domain of human geography" in *Directions in Geography* Ed. R Chorley (Methuen, London) pp 67–87

Hägerstrand T, 1974 "Comment" in *Values in Geography* by A Buttimer (Association of American Geographers, Washington, DC)

Hägerstrand T, 1978 "Survival and arena" in *Timing Space and Spacing Time: Human Activity and Time Geography* Eds T Carlstein, D Parkes, N Thrift (Edward Arnold, London) pp 122–145

Haggett P, 1965 *Locational Analysis in Human Geography* (Edward Arnold, London)

Hartshorne R, 1939 *The Nature of Geography* (Association of American Geographers, Lancaster, Pa)

Hartshorne R, 1955 "'Exceptionalism in geography' re-examined" *Annals of the Association of American Geographers* **43** 226–249

Harvey D, 1969 *Explanation in Geography* (Edward Arnold, London)

Harvey D, 1973 *Social Justice and the City* (Edward Arnold, London)

Hindess P, Hirst P, 1977 *Mode of Production and Social Formation* (Macmillan, London)

Hurst M E, 1980 "Geography, social science and society: towards a de-definition" *Australian Geographical Studies* **18** 3–20

Jensen-Butler C, 1981 "A critique of behavioral geography: an epistemological analysis of cognitive mapping and of Hägerstrand's time–space model" DP-12, Geographical Institute, Århus University, Denmark

Johnston R, 1980 "On the nature of explanation in human geography" *Transactions of the Institute of British Geographers* **5** 391–412

Layder D, 1981 *Structure, Interaction and Social Theory* (Routledge and Kegan Paul, Henley-on-Thames, Oxon)

Lemert C, 1979 *Sociology and the Twilight of Man* (Southern Illinois University Press, Carbondale, Ill.)

Lévi-Strauss C, 1966 *The Savage Mind* (Weidenfeld and Nicolson, London)

Lévi-Strauss C, 1969 *The Elementary Structures of Kinship* (Eyre and Spottiswoode, London)

Lipietz A, 1977 *Le Capital et Son Espace* (Maspero, Paris)

Mårtensson S, 1979 *On the Formation of Biographies in Space-Time Environments* (Gleerup, Lund)

Marx K, 1976 *Capital: A Critique of Political Economy* (Penguin Books, Harmondsworth, Middx)

Olsson G, 1980 *Birds in Egg/Eggs in Bird* (Pion, London)

Peet R, 1978 "Materialism, social formation and socio-spatial relations: an essay in Marxist geography" *Cahiers de Géographie du Quebec* **22** 147-157

Poster M, 1979 *Sartre's Marxism* (Pluto Press, London)

Pred A, 1977 "The choreography of existence: comments on Hägerstrand's time-geography" *Economic Geography* **53** 207-221

Pred A, 1981 "Social reproduction and the time-geography of everyday life" *Geografiska Annaler* **63B** 5-22

Robson B, 1973 *Urban Growth: An Approach* (Methuen, London)

Rose C, 1977 "Reflections on the notion of time incorporated in Hägerstrand's time-geographic model of society" *Tijdschrift voor Economische en Sociale Geografie* **68** 43-50

Rubinstein D, 1981 *Marx and Wittgenstein: Social Praxis and Social Explanation* (Routledge and Kegan Paul, Henley-on-Thames, Oxon)

Sack R, 1972 "Geography, geometry and explanation" *Annals of the Association of American Geographers* **62** 61-78

Sack R, 1974 "The spatial separatist theme in geography" *Economic Geography* **50** 1-19

Santos M, 1977 "Society and space: social formation as theory and method" *Antipode* **9** 3-13

Sartre J, 1976 *Critique of Dialectical Reason* (New Left Books, London)

Sayer R, 1979 "Epistemology and conceptions of people and nature in geography" *Geoforum* **10** 19-44

Schaefer F, 1953 "Exceptionalism in geography: a methodological examination" *Annals of the Association of American Geographers* **43** 226-249

Sheridan A, 1980 *Michel Foucault: The Will to Truth* (Tavistock Publications, London)

Smith N, 1980 "Symptomatic silence in Althusser: the concept of nature and the unity of science" *Science and Society* **44** 58-81

Smith N, 1981 "Degeneracy in theory and practice: spatial interactionism and rdcial eclecticism" *Progress in Human Geography* **5** 111-118

Smith N, O'Keefe P, 1980 "Geography, Marx and the concept of nature" *Antipode* **12** 30-39

Soja E, 1980 "The socio-spatial dialectic" *Annals of the Association of American Geographers* **70** 207-225

Soja E, Hadjimichalis C, 1979 "Between geographical materialism and spatial fetishism: some observations on the development of Marxist spatial analysis" *Antipode* **11** 3-12

Steiner G, 1975 *After Babel* (Oxford University Press, London)

Steiner G, 1981 *The Portage to San Christobel of A.H.* (Faber and Faber, London)

Tuan Y, 1977 *Space and Place* (Edward Arnold, London)

Urry J, 1981 *The Anatomy of Capitalist Societies: The Economy, Civil Society and the State* (Macmillan, London)

van Paassen C, 1976 "Human geography in terms of existential anthropology" *Tijdschrift voor Economische en Sociale Geografie* **67** 324-341

Vilar P, 1973 "Histoire marxiste, histoire en construction: essai de dialogue avec Althusser" *Annales: Economie, Société, Civilisation* **28** 165-198

Williams R, 1980 *Problems in Materialism and Culture* (Oxford University Press, London)

Williams S, 1981 "Realism and the objects of human geographical inquiry" *Area* **5** in the press

Wood P, 1978 "Location theory and spatial analysis" *Progress in Human Geography* **2** 518-525

Languages and dialectics

–/–

Gunnar Olsson¶

To translate is to express a sense in another language-parole. It is to carry to heaven without death and to remove the dead body or remains of a saint. It is to convey an idea from one art form to another. It is to make new boots from the remains of old ones. It is to interpret sings.

The definitions come from the OED. The conclusions are my own: Much is aVOIDable, translation is not. Anywhere/anytime, aeneymy/anyf(r)iend.

Anyhow:

My · is made, now I must erase it. Thus, to translate is to doubly lie, to beget by not getting at the truth. Little wonder then that all social laws are laws of the double. More wonder that most social scientists are one-eyed cyclops unable to imagine perspective; misled by our singular vision we tend to confuse use and mention, word and object.

Where is he, that wondrous wandering wonderer capable of releasing Ulysses from his wake?

* * *

In the convent, the point is more conventional:

All understanding involves crucial elements of translation, of movement from one conceptual world to another. Thus, all understanding is by necessity metaphoric, for I must always grasp what I wonder about as something different; the I becomes an Other, the Other becomes a Me.

It cannot be said more clearly:

To wonder about understanding is to be involved in language. But, to be involved in language is to exploit the distinctive connection between name and object, thing and relation, appearance and essence.

Already Odysseus caught in the Cyclop's cave knew that there is a distinction between what I think-and-say and what I think-and-say about. Wondering about understanding must consequently not be limited to the semantics of the signs which signify what I am talking *about*. It must also involve the pragmatics and syntax of the categorizations and relations I am talking *in*. The challenge is not to be tuned to the vibrations of my vocal chords or to detect the spots I leave on the white sheet. It is rather to be aware of the silent forces of that subtext which sneaks away into the emptiness that ties the marks together. But the overwhelming practice is to concentrate solely on what appears on the lines. The blank spaces which separate and unite them go unnoticed, precisely because they separate and unite.

¶ The author is grateful to Allan Pred for conversations, Bourdeaux, pheasant paté and Brie.

So:
See not only what is on the lines, but
also what is between them.
Read less of what I am sufficiently ignorant to write and more of what
you know so well that it must be passed over in silence! Deafen yourself
to the noise of the expressible! Listen instead for the whisper of the
taken-for-granted! But be most curious about the limits between
categories, for it is only in the act of crossing a boundary that you
mistranslate and consequently learn! Everything else is obedient
reproduction.

* * *

Virtually everything is reproduction. But virtue is in the constraining
constraints of the mimicking social scienses, vice in the liberating
possibilities of the creative arts. This makes it possible to predict the
future (as in Foucault's scientia sexualis) and impossible to presense it (as
in his ars erotica). And yet, it is part of logic itself that 1984 is in the
midst of the 1980s. I therefore write now as a way of anticipating this
new world at the decade's end.

The outlines of this new world are already present, for how could it
otherwise be recognized. It is a world in which the familiar industrial
mode of production is overtaken by a hitherto unknown state mode of
production. Although the transformation can be read off the patience
cards that already lie on the table, this is not to say that the present
determines the future. Rather it is often the reverse, because our hopes
and fears for the future let us see only certain aspects of the present.
But what is shown to the observer are merely masks. The future itself
remains as invisible as the ontological stuff it is made of.

The following are nevertheless 1990-oriented questions to the 1980s:

Which aspects of the present are taboo because they are too important
to reveal? Which types of legitimating unknowing is society's science in
the midst of creating? How is today's state-capitalism disguising its
fundamental contradictions? How do I notice and then interpret the
signs of the rainbow in the sky? Who understands the silent language of
the taken-for-granted well enough to translate it? How can I re-member
what others have forgotten and how can I forget what others de-member?
Is the trancelation of relations the deconstructivist's version of
Wittgenstein's throwaway ladder?

And then:

Is it too early to translate the language of state-capitalism or too late
to trance-end its imprisonment?

* * *

Anyone who understands me eventually recognizes these stripteasing
questions as nonsensical, when he has used them—as steps—to climb up
beyond them.

He must transcend these rhetorical propositions. But he will not then,
as Wittgenstein suggested in *Tractatus* 6.54, see the world aright. He will

instead have created another world surrounded by other limits, guarded by other silences, ruled by other emperors.

How will the rulers be dressed tomorrow?

* * *

Just as noone can fully grasp Marx's *Capital* without first having studied through and understood the whole of Hegel's *Logic*, so noone can fully understand state-capitalism without first having internalized the meaning of the sign /. This sign is a symbol of relations, of the unity between identity and difference.

In conventional reasoning, relations are not denoted by a slanted line but rather by the parallel lines of the equality sign. This sing is then interpreted in the Leibnizian spirit of salva veritatae and the Russellian matter of logical atomism; a proposition is held to be both true and informative only if the equality sign is flanked by a proper name on the one side and a definite description on the other or, alternatively, by two different definite descriptions of the same object. But even though the whole point of

$$E!(\iota x)(Qx) = (\exists b)(x)[(Qx) \equiv (x=b)]$$

and

$$U(\iota x)(Qx) = (\exists b)\{(x)[(Qx) \equiv (x=b)]\&(Ub)\}$$

is to define away the definite description $(\iota x)(Qx)$, this *Principia* trick (*14.02 and *14.01) of abolishing definite descriptions by not mentioning them does not alter the fact that understanding requires their *use*. In use, however, definite descriptions reveal themselves as what they really are: contextual, metaphoric and self-referential. This is indeed why Russell wanted to rid them from all analysis; in his own words, "every proposition and every belief must have an object other than itself".

Materialists must now ask the idealist question:

Which is the object of state-capitalism and its precursing postmodernism, if it has to be an object other than itself?

The emerging answer:

Perhaps there are no such objects, for the major characteristic of state-capitalism is that it is paradoxically locked into itself.

Next question:

How is it constructed, the self-referential reasoning net capable of catching the emerging world of masturbation?

* * *

By writing / instead of =, I signal my interest in dialectical, internal and self-referential relations, names that most analysts have learned neither to use nor to mention. The words themselves are banned by the church of Fundamenalists.

The ban was issued because all relations (including equalities) are of an ontological kind alien to the ruling ideology of presence. Thus, relations are by necessity invisible as the emptiness between the lines of my text, inaudible as the silences that turn meaningless noise into meaningful words; the untouchable is pariah, the pariah untouchable. In society's interest of communication, there is consequently a strong tendency to do away with relations by thingifying them. In this tragedy of the common, however, the interest is turned from the concept of the relation itself to the phenomena or things related. Rather than questioning the relations we are talking *in*, we stare ourselves blind on what we are talking *about*. Instead of wondering about = or /, we get caught in the sign – –. This sign is a symbol of things related.

Direct contact with / is culturally forbidden. This is why it intrigues me, for whatever is dangerous enough to be taboo is important enough to understand. The constructive is not to question the static means of representative samples but to unravel the dynamic variances of distributional tails. And yet, culture is founded on its limits, civilization on its madness.

Thus:

Relations are not only relations between measurable things but also between cultural words, not only words but concepts, not only concepts but meanings, not only meanings but other relations. It follows that relations are always related to other relations all connected into strangely looping spirals. Self-reference is the word for this peculiar concept coiled at the center of current thought and extending beyond its frontiers.

Self-reference is the key to the coming revolution of the social sciences. Where are the lock-smiths who know the code?

* * *

Relations like beauty, sincerity, trust, malice, disgust, and nausea are not in the things themselves but in culturally determined conceptions and behaviors; isolated things are as meaningless as connected relations are meaningful. This is why comparative studies are both so promising and so dangerous; promising because they lead to understanding of the I through the Other, dangerous because they are potentially emancipatory. Benjamin, Horkheimer, Adorno, and Marcuse all set examples. But so does everyone else who in exile experiences how one never learns home until one goes away. But each exit is an exit with no return as Homeland becomes the safe symbol of escape.

Illustrative examples are in Geertz's analyses of the Balinese cock fights. For a western anthropologist it is easy to see those ritual dances as being performed not by roused birds but by people who via their cocks tie and untie knots of family relations. It is more difficult to interpret your reading of this writing as an integral part of geography's death and initiation rites.

And yet, a rite of author(ity) is exactly what this dual relation is: Preparations for the cooking of the raw. But who knows the recipe?

Who are the cooks and who are the cooked? And who furnish the pot and the tempting spices?

* * *

The relations which tie the one to the many and the many to the one are at the same time determining culture and determined by culture.

Writing those words is easy. Reading them is incredibly difficult. The reason is that whenever I talk *about* culture I must talk *in* culture; culture is like its own language in that it is bound to use itself to understand itself. And so it is that any social scientist is handicapped by the methodological praxis which requires him to be more stupid than he actually is. Thus, in the interests of discipline, verification, and communication he relies mainly on the two senses of sight and hearing: What counts is what can be counted; what can be counted is what can be pointed to; what can be pointed to is what can be unequivocally named. Accumulation of knowledge about the nameable is consequently the point of the scientist's game. Power, though, is not in uttering the nameable things of commodity fetishism and penis envy but in innering their symbolic condensation of relations: Un Coup de Dés played with loaded dice.

Imagine here a Foucaultian study of filth and human excrement. To see and hear the shit is barely passable. Uh! To touch it is nauseating. Woh! But smelling and tasting it penetrates so deeply into the ego itself that it is almost unthinkable. And so it is that killing a thousand people by target-seeking robots is acceptable. But killing one person with a bloody throat-bite is so brutish that the thought itself takes its holder to the asylum.

* * *

The interesting is not to note obvious facts of empirical behavior. It is rather to wonder about the particular socialization processes whereby individual and society are brought together. But this is to wonder about the taboos associated with the limit between the Ego and the Other. Put differently, the issue is how you and I distinguish ourselves from each other by establishing impregnable boundaries between us.

Perhaps the question is:

If definitions require distinctions, do relations affirm them by transcending them?

Or, more operationally:

How do I teach my children to tell the truth and yet realize that truth telling can sometimes be evil and therefore forbidden? How does the Family Circ(l)e turn human beings into swine?

* * *

Relations are often called mystical and thereby silenced. Perhaps this practice reflects the fact that proper understanding of social relations is a prerequisite for the understanding of power. But such an understanding

is too fundamental for society to afford. Adam's apple is a double
symbol of temptation-and-fall and of knowledge stuck in manly throats.

Power is another word for the relation between the I and the non-I.
It follows that the process of liberation can never end, for its driving force
is in the emancipation and creation of the self. Once this is understood,
it is easier to see not only that all power involves issues of translation but
also that every power struggle is a struggle of independence; power
would not be power if it were not a relation, that is, if it were not of an
ontological kind different from the things in which it momentarily seeks
to hide. He who has power knows how to mislead by mixing ontologies,
pretending to be concerned with things while in reality knowing that
things are meaningless until tied into meaningful relations. The essence
of power is thus in the slanted /, its appearance in the repetitive – –.

And so it is an integral part of all relations that we tend not to notice
them until they begin to malfunction. Neither do I notice the air I
breathe or the blood in my veins until the relation between them is
disturbed. When it is, however, then every doctor is preconditioned not
to wonder about the relation *per se* but rather about the things of oxygen
and pumping muscles; we ask not about the relation /, but about the
categories – –.

In this movement from / to – –, questions of epistemology turn to
questions of ontology:

Is the practice of defining our problems into existence a technique for
getting at truth or for defining them away? Is it the practice of those in
power to thingify relations and thereby block the road to deeper insights?
Is it in society's collective interest to mislead its individuals into seeing
appearance rather than essence? Are we stuck in the serpent's truth that
"God knows that when you eat of the fruit of the tree in the midst of
the garden your eyes will be opened and you will be like God, knowing
good and evil"? Is the sign "God" nothing but the proper name of the
definite description "the collective unconscious"? Is the Barefoot-Father-
with-the-Beard a fetish of that social glue which is important enough to
be taboo? Is the crucified son of flesh and bone merely another stage-
stop on the ontological journey from subsisting relations to existing
things? Is reification deification, deification reification?

Yes!

The reason for the yes is that there is no objective reality to reflect
upon, for what appears is itself essentially a reflection of the reflector's
subjective self-awareness of that reality. As the dialectics of flexuose
flexion runs its course, questions of ontology therefore turn to self-
reflective questions of mythology:

Why is the serpent the symbol of self-reference? Is it because it knows
the secret of the collective unconscious?

* * *

When a social scientist deifies by reifying, he christens the sign – – as "society and individual". There is much to indicate that the dialectical interplay between these two categories currently is under serious strain. Perhaps the malfunctioning is most illustrative in welfare states like Sweden, where the crisis is less a matter of resources and more one of demoncracy. For what other is democrary than a powerful set of principles whereby one-and-many, many-and-one are forged together into what is presented as functional efficiency and moral justice? Put differently, the principles of demoncrazy shape and reflect how the psychological concept of the ego is translated into constitutional law; the high court positioned as frontier guard in the wasteland between I and the Other, therefrom ruling over what is equal to what, over good and evil, life and death.

Some claim that the social collective now has penetrated deeply into the realm of the individual. Others note the complementary trends toward privatization. In my interpretation, these same tendencies are further indications that capitalism is in the midst of a rapid, decisive, and irreversible transformation from industrial capitalism into a form of centralized state-capitalism. Both individual and society have yet to adjust to this fundamental change from one dominant production mode to another. For this to occur it is necessary to develop new decision procedures, perhaps even new personality types. Whether we like it or not, that is also the direction in which we are heading. As in the previous shift from the feudal to the capitalist mode, devils and witches are invented as scapegoats for commodity fetishists.

Who are the witches today? How is it determined whether they sink or float?

* * *

Now it can be thought-and-said:

There is time for a new Marx. This is due both to the atrocities committed in his name and to the disrepute brought by his Parties to dialectics. But it is mainly because objective reality no longer is the same; as Marx himself foresaw, quantitative changes in capital have led to qualitative changes in Capital. It is obvious, for instance, that the modern suburbs of Stockholm differ drastically from those of Manchester a century ago, not only in their outer form but also in their functioning. And yet, the sense of human deprivation inherent in the repressive domination of man by man may be just as intense now as then. Thus, the human sacrifices continue, for the gods of social cohesion and rational exchange have simply changed from the suit of the old capitalist into the open shirt of the new social bureaucrat; what is inside the velour pants is nevertheless the same as was inside the strip(p)ed trousers. Perhaps most analysts were too busy trying to understand the old world to notice how the new was changing. Perhaps he who once was turned on his head now is being turned back on his feet.

The emerging / in individual/society seems most evident in the grass-root movements currently spreading throughout the developed world. Here it is striking how the protests now focus less on the conditions of work and more on the holy family itself; it is in the micropowers of daily life at home, nursery, school, commuting, and hospital that we concretely experience the modern forms of social imprisonment. It is in those spheres of immediate existence that society reveals its fundamental contradictions of unfairness. It is through changes in familial relations that state-capitalism both reveals and hides itself.

Timely questions:

Which identity crises are in the commodities of the culture industry? Who would Oedipus have been without Laios? How do you rid yourself of the superego, if the superego is not a person but a faceless collectivity? Where does it reside, the fearful authoritarianism of state-capitalism?

* * *

It is easy to turn to Habermas on the ensuing crisis of legitimation. It is nevertheless more important to wonder about counterfinality, that is, about how we came to live in a world that is opposite to the good intentions it grew out of. But counterfinality flies in the face of traditional thought, for it is a situation in which the truths of the premises have not been preserved in the conclusions. It raises the issue of how I tell truths about a world whose very nature it is to be a lie:

What reasoning tools do I employ when I realize that the social world does not obey the rules of conventional truth-functional logic? What do I do when I notice that state-capitalism has many traits in common with such enemies of our culture as paradox and tragedy? Who is to blame when everything is perfectly right in the beginning and everything horribly wrong in the end?

Noone is to blame, for noone has broken those behavioral rules of reasoning into which he has been socialized. The tragic hero as expression of the Eros of the western ethos!

So:

How are we socialized into thinking-and-acting in ways which are at the same time individually praiseworthy and in the interest of society's state-capitalism? How do I move from categorizing crosses and directional arrows to self-referential loops? How do people learn to live in institutionalized double bind without going crazy?

* * *

The transformation of industrial capitalism into state-capitalism is already evident in the socialization and reproduction processes whereby society and individual are being adjusted to each other. By fulfilling moral codes, we experience how the temptation to dream and transform is overcome. But to experience is not to understand, even though it is a necessary step toward that boundary between the I and the Other where understanding resides. This boundary is taboo, now as much as in Paradise itself. In the process of trespassing, the anxiety of relations is turned into the fear of things, issues of power into fig leaves. And yet, castration is the metaphor that (fe)male power seems most eager to suppress.

So:

How can I simultaneously anchor – – in / and / in – –? If I ever did, how would I then translate my insights into communicable expressions without destroying them? How can you and I as individuals eavesdrop on society when it thinks-and-talks about itself in-through itself? How do I capture the dialectic of society and individual without falling into the – – trap of sociology and psychoanalysis?

* * *

Caught in culture, the only way to produce is to reproduce by putting words out of conventional contexts, by making new boots from the remains of old ones. Thus it is in self-reflection that reason sees its own interest. To ask again is consequently not to repeat but to translate anew.

Therefore:

Which forces of social cohesion are illustrated and further entrenched in the Odyssean act of ontological juggling? Is his appearance essential or is his function merely to divert attention from the pickpockets that raid the appalauding audience? How is this desiring piece of writing itself a legitimating instance of the socialization processes of state-capitalism? Is not striptease the appropriate metaphor of a society that talks about its own silences, reveals the powers it exerts, and promises to rid itself of the laws that protect it?

What does it mean to engage in dialectically mediating history-specific communication?

* * *

The challenge is enormous. Not for geography or any other well ordered discipline, but for its individual members exploring the limits of culture. Mallarmé was eons ahead of Christaller:

NOTHING WILL HAVE TAKEN PLACE EXCEPT THE PLACE EXEPTÉ

PEUT-ÊTRE

UNE CONSTELLATION

Toute Pensée émet un Coup de Dés
Hazerdous hazard.

Dialectical analysis of value: the example of Los Angeles

Bernard Marchand

The adaptation of housing to human needs

In Los Angeles, as in many other great cities of the world, population and housing have basically changed in two ways. First, we can identify concentric waves of diffusion moving through the urban space at varying speeds, waves of building and change that are often related to new forms of transportation. Second, there are clear sectoral alignments, areas characterized by different types of inhabitants who change from one time period to another. These rather fundamental differences in spatial structure, evolutionary rhythms, and urban relations pose some difficult questions of human dynamics in urban areas. For example, how do residences adjust over time to the requirements and resources of their inhabitants? What forms do such adjustment take, and how do they evolve?

As we examine the historical development of any urban area, we face essentially three questions:

1. How were the needs of everyday life (for example, housing and work) satisfied? The main criteria here seem to be such things as dwelling size and type, as well as the degree of accessibility to work.
2. What were the relationships between housing costs and the ability of households to raise funds for housing? This question focuses mainly on the structure of mortgage financing.
3. How did the population accommodate itself to an evolving pattern of residences and changing family requirements? In many cities, certain evolutionary changes have been 'out of phase': these sorts of problems have been particularly marked in ghetto areas, as the population expanded at the same time that the buildings became more dilapidated. The marked contrast between housing characteristics and actual family needs is part of a larger problem—the overall evolution of spatial structure in the city.

In actual fact, each of these questions refers to an even more basic concept: namely, what is the *value* of a dwelling for its inhabitants? How does a dwelling evolve through time, and how are various qualities of residences distributed over the urban space? In the context of this essay, the very ambiguity of the concept of *value* is useful, since it is a term that covers a complex reality which may be analyzed at a variety of different levels (Baudrillard, 1972). Although I shall use Baudrillard's categories in the course of this analysis, I do not wish to imply that I entirely accept his theses, nor do I wish to follow his form of analysis step by step. Nevertheless, his classification is a convenient one, although we must remember that it is still provisional and lacking somewhat in precision.

Housing values

Baudrillard defines four types of value which may well be different for the same dwelling (table 1):

Use Value. This is the classical concept, representing the usefulness of an object in everyday life. As far as a single dwelling goes, its use value varies according to the needs of the inhabitants. For example, the large apartments in the center of Paris, which were once luxurious but are now dilapidated, have been increasingly occupied by poorer families with many children. These people have no use for monumental doors or high ceilings, but suffer from the small number of bathrooms and other sorts of facilities. The use value of these apartments tends to decrease with time —a characteristic process in the central areas of many cities.

Exchange Value. This appears through equating different dwellings which have the same commodity or monetary value expressed in the housing market. It normally depends on the state of the market, on the equilibrium conditions holding between supply and demand, and on numerous factors which are quite independent of the dwelling itself: for example the degree of accessibility, the quality of the neighborhood, the prestige of the area, and so on.

Sign Value. This arises in urban areas from a direct transcription of concepts that are essentially linguistic. Every residence—its location, style, use of space, and so on—bears witness to the social position of the owner or the tenant. A major goal is to differentiate one person from another, often by using a kind of social language written in stone or concrete. Here the role of the sign value is diametrically opposed to the role of the exchange value. Instead of establishing an equivalence between different objects, sign value tries to create and express differences between objects that are actually perceived as *too* similar. The best example might be the plaster frontages, so common in Los Angeles, which outwardly offer the most varied forms and colors, but in fact conceal miles and miles of similar wooden frames (Banham, 1971, pages 173, 197). In a sense, architectural details are the *vocabulary* of this language, while the spatial organization of the buildings, the gardens, the fences, the trees, the parking lots, and so on form its *syntax*.

Symbol Value. This concept concerns what we might call the deeper Freudian categories as they relate to architectural forms. We can think of these as acting unconsciously on the human mind to explain why some people are mysteriously attracted to some places and buildings, while they may be completely indifferent, or even repulsed, by others. Feelings are often ambivalent and contradictory, as they so often are in the Freudian context, because they are essentially prelogical. They do not obey the principle of noncontradiction—as Kevin Lynch demonstrated implicitly in

Table 1. The major sources of housing value[a].

Value	Basic principle	Role	Nature	Domain of realization
Use	uniqueness of the home (*The One*)	fulfilling a function	a purely *concrete thing*	in everyday life
Exchange	equivalence of homes (*The Many*)	as money	a purely *abstract thing*	on the land market
Sign	communication between *many* people	as language	a purely *concrete idea*	in social relations
Symbol	an *individual's* unconscious life	as satisfaction of one's unconscious drives	a purely *abstract idea*	in individual affective life

Value	Logic	Dialectical realization	Example: A garden as a ...
Use	Qualitative	the object is used, that is, it is consumed and destroyed (in this case, as the dwelling is occupied and so is not useful for others)	children's playground
Exchange	Quantitative	the dwelling is transformed into something else, that is, money, which is itself nothing but the potential for objects (and dwellings) to transform themselves	local advantage increasing the market price
Sign	Differentiation	the dwelling describes its occupants, but only in order to separate him, to differentiate him from other city-dwellers	witness to an occupant's social position in relation to others
Symbol	Ambivalent (the noncontradiction principle does not apply here)	the unconscious effect of the symbol is prelogical and ambivalent; that is, it is made up of contradictory drives (love and hate, desire and disgust, attraction and repulsion, simultaneously)	evocation of a wilderness, attracting and frightening at the same time

[a] Inspired by Baudrillard (1972), but largely modified.

his work on the perception of the urban landscape, particularly in Los Angeles (Lynch, 1960). Bachelard (1957) has also insisted upon the usefulness of such an approach, with many convincing examples chosen from European literary texts.

The first two values above might be labelled *classical*, and they tend to feed the normal sort of polemic that has been going on in economics for the last 150 years. But the last two appear to hold much more promise, although they have not been widely used in urban analysis. The townscape of Los Angeles appears as a particularly appropriate subject for an analysis based on these latter categories. Of course, it can be argued that the exchange value somehow synthesizes and represents all these other concepts of value. However, it seems more reasonable to integrate these four concepts in the notion of price. I shall not try here to analyze the relationship between price and value: this old problem has little bearing on this particular essay.

The dialectics of value

The four basic categories of value, together with what we might call their 'inner mechanisms', are related in dialectical ways. The dialectical approach focuses upon the development of compatible and opposing relations between the three logical values, while the fourth one (symbolic value) contains a built-in dialectic of its own—since it is, in fact, prelogical. As is characteristic of an analysis of this sort, each value realizes itself by destroying itself, thus exhibiting its true dialectical nature.

As we approach the analysis of an urban area within this framework, it should come as no surprise that we recognize first the opposition between the One and the Many, a very old dialectical relation that leads us straight back to Plato's *Parmenides* (table 2). Use and symbol values have meaning only for the *individual*: an object (particularly a house or an apartment) fulfills quite different and distinct functions for each person, while symbols arouse unconscious impulses of an infinitely varied nature in each of us, often related to the most intimate memories of our early years. Thus, the values of these categories change not only from one person to another, but also for a particular individual at various stages of his or her life.

The *use value* depends on the most concrete apprehension of a *thing*: for example, how do the material qualities of an apartment—its size, form, inner layout and arrangement—fit our particular needs? In contrast, the *symbolic value* relies on the abstract effects of *ideas*—the images, memories, and desires they generate—and its domain lies at a deeper layer of our psyche, so remote from the world of action we cannot investigate it at will or act upon it, but so important that it determines each of our acts. Both of these values are completely opposed both in their form and in their domain, and yet, at the same time, they are completely similar inasmuch as each one determines the behavior of an individual—in a sense, one acting from the outside, while the other works from the inside.

The thing that these two opposed values have in common is that they are both defined for a particular individual, and at the same time define that person in turn. In marked contrast, *exchange value* and *sign value* have meaning only for a collection of people. Both of these values serve to relate an object (and the person behind the object) to another: a commodity like a house is related to another commodity like a sum of money by indicating their equivalence, or exchange, value. Similarly, an architectural or urban form (for example, a form of decoration or layout), is related to another form by demonstrating the value of their difference or sign value. Both of these are similar on different levels, even as they play opposite roles in the same collective or social context.

We know from Hegel that the One and the Many have no meaning when they are artificially separated from one another, and that they actually merge incessantly as each disappears into its opposite to create *change*—which is their actual form of existence. This logical dynamic suggests an interesting relation between a dwelling's *use* and its *symbolic* value: the experiences we have had of our home spaces generate the bases for the symbols which affect us. Conversely, these symbols from the unconscious shape our perception of space, and determine how we actually use our dwellings. This continuous and complicated dialectic over the course of people's lives is expressed in the concept of *neighborhood*, a piece of urban space defined by its inhabitants who present a certain degree of unity and share a number of similarities. However, the concept of neighborhood begins to take on a collective form, which then feeds back to influence the individuals who make it up, molding their tastes, their feelings, and their behavior—particularly the way they bring up their children. Thus the neighborhood should not

Table 2. Dialectical relations between values.

Cross-relations	Thing	Idea
Concrete	Use	Sign
Abstract	Exchange	Symbol

Main diagonal — *The individual life*
Values of opposite natures with similar roles: Use ↘ (One) ↘ Symbol

Secondary diagonal — *The collective life*
Values of opposite roles with similar natures: Sign ↙ (Many) ↙ Exchange

only be considered as the sum of the individuals making it up, but also a collective form acting on the individuals, so ensuring the social reproduction of the group itself. Its highly dynamic character is obvious, as it becomes a necessary framework for individual and collective change.

Exchange value is a purely *quantitative* concept, based on the possibility of expressing any object in terms of something different from itself (whether it be cattle, coffee, silver, gold, or any other form of money) by establishing an equivalence between the two things. We say that an object is worth so many currency units, and that these express in turn the worth of another object. The relation is reflexive, symmetrical, and transitive: it is precisely an equivalence relation. At the same time, it negates any qualitative difference between objects, which are now perceived as though they had no particularities of their own. Thus exchange value, emphasizing the purely quantitative aspect, is the very negation of the idea of quality and qualitative difference.

Use value forms the exact antithesis of exchange value. It is based on all the individual peculiarities of an object as it is compared and contrasted with another. From the viewpoint of use value, all objects, whether commodities or services, are unique and can never be equivalent. The concept here is entirely *qualitative.*

Sign value is somewhat more complex: by expressing in an architectural language the wealth, taste, and upbringing of a landlord or the occupant of a building, sign value 'tells' us about social quality as expressed by quantity. As such, sign value is nothing but the *synthesis* between use and exchange value, as it retains both of these elements in dialectical opposition, and transcends them into a new equilibrium—namely, money. Money (that is, income) is expressed by the sheer size, or the number and the kind of decorative elements on a house (to focus only on one or two possibilities), but it is not simply a display of money. Taste, in the sense of whatever judgement we may pass on it, modifies and contradicts the sheer weight of money. In the same way that poverty may sometimes try to conceal itself by exaggerating stucco decor, so wealth might be delicately concealed in architectural understatement. When an owner 'says' something in terms of the size, shape, and style of his house, it is through the complicated and dialectical interplay between the brutal quantity of dollars invested in it, and the quality of taste which uses such financial raw material, molding it into an expression of the best quality—the word "best" referring, of course, to the inner personal judgement of the particular person.

In this way, sign value simultaneously integrates, combines, and negates both exchange and use values. In fact, the two elementary values have no real meaning if they are separated from each other. For example, in most of Western society how could we assess the exchange value of a dwelling on the market divorced from any reference to its use capabilities?

Conversely, what is the point of discussing the use of a certain type of building without indicating either its profitability or the kind of occupant who will be able to afford it? Sign value integrates both of these basic concepts, and for this reason it is probably much closer to actual market price.

Symbolic value stays outside of this dialectical trinity. It is a prelogical form of thinking, taking its cue from the working of the Freudian Id, and not from the conscious Ego. Typically it is ambivalent, like those unconscious impulses which unite in the same drives the feelings of love and hate, desire and disgust, presence and absence. Its very nature is dialectical. While the three preceeding values interact in the conscious mind, symbolic value constitutes their negation—even as it is simultaneously embedded in each of them. It is present in exchange value, as it contributes to the determination of the price one is willing to pay by adding attractive or repulsive constituents to the decision process. It is present in use value, since it adds a whole galaxy of feelings to the practical aspects of a dwelling. Finally, it is present in sign value, since it imbues nearly all signs of wealth, success, culture, and taste with unconscious 'connotations' which radically change the meaning of an architectural message. By its presence in sign value, language is altered into myth: for example, the use of phallic decoration as expressions of power and domination, seen in the huge chimneys sticking out of the roofs of so many Beverly Hill mansions, when the inhabitants think they only want to say "Home Sweet Home"! Sign value embodies and unites exchange and use value in a contradictory way; symbolic value is the negation of all three. The final relationship boils down to the contradictory connection between *language* (sign) and *myth* (symbol), both of them similar in the sense that they express ideas, but both very different and opposite in the way they express them.

What we must understand is that each value represents some potential for a particular object, a potential for being used for something, for being exchanged against something else, for expressing some social message to the onlooker and arousing feelings in him. These potentialities are realized in a dialectical way, by a process of self-negation that turns an object into something else, so changing its nature and its purpose in turn (table 1).

We have already seen how the first two values operate in this process of negation. In using an object we actually consume and destroy it. In the case of a dwelling, this can take two forms. On the one hand, the dwelling needs regular maintenance, repainting, and so on; on the other, it represents a potential home for anyone, but once an occupant has actually settled down it cannot be used by anyone else. Dialectical transformation is still more obvious in the case of exchange value; the physical structure of the dwelling has to be transformed into an opposite form of capital, namely, money, in order for its value to be realized.

In a similar way, a dwelling describes, through the public message of its sign value, the social position of its occupant in order to distinguish that person from all other city dwellers. The point here is not to decide if this value is high or low, or even if the message is clearly understood, but to clarify the *way* in which it actually works.

As I noted before, symbolic value follows a rather different and original process, not being subject to the principle of noncontradiction, but rather to a powerful and prelogical mechanism of association of ideas as in dreams or cases of *lapsus linguae*. Symbols express contradictory feelings which have not been clarified and sorted out, but coexist together albeit in opposition within the symbolic object.

It is quite possible that further analysis along these lines may lead urban planners and sociologists to refine these concepts of values, or even redefine them. Nevertheless, as they have been outlined by Baudrillard and redefined here, they seem to offer a most useful and interesting platform for urban analysis. By their nature, their evolution, and their internal relationships, they are totally dialectical and dynamic, constituting the very categories through which a city changes.

The nature and evolution of sign value

Sign value has, in essence, a linguistic function; namely, to express and 'tell' differences, to transform locations into social signs, and to constitute, through the spatial combination of such signs, an urban language. We are all familiar with signs of well-being expressed by the differential desirability of locations. The names of the pleasant districts built on the hills between Santa Monica and Hollywood are signs of opulence and well-being, and their high land values partly include such signs of success. This part of the Los Angeles urban region has retained its privileged meaning for at least thirty years—as a survey made by geographers at UCLA in the 1970s indicated (Clark and Cadwallader, 1973). Asked which district they would like to live in *given their present income*, one thousand Angelenos designated (1) the hill slopes to the northwest of the city, and (2) the ocean side districts of Redondo, Manhattan, and Hermosa Beach. There is, of course, a use value included in their choices—the presence of nice views, not too much pollution, and so on—but there is also the prestige of the names constituting the sign values of these residential districts. We only have to think of Pasadena: it appeared high in the list of most desired locations, even though its townscape is visually mediocre, and its pollution rate extremely high. Perloff, intrigued by such a high evaluation, noted that "Pasadena's choice seems to be related to its preeminence as a residential zone forty years ago" (Perloff et al, 1973, page 206), indicating that its sign value is still extremely high.

The importance of sign appears in the very names of the most desirable locations. In the hill areas we have all sorts of Crests, Hills, Glens, and Canyons signifying relief (in the double sense of the word), domination

over the environment, aloofness from other city-dwellers, as well as
seclusion, quietness, and peace. Similarly, along the shore we find such
terms as Beach, Marina, and Ocean expressing a brisk and healthy contact
with the sea, surf, sand, and sun, the vastness of the Pacific, and the sense of
openness to the exotic countries beyond it.

Modern linguists, particularly Roland Barthes, have shown that language
is not only spoken. Many parallel languages—such as fashion (Barthes,
1967), body movements and objects (Baudrillard, 1968)—can tell us as
much, and often more, than the usual oral discourses. Inasmuch as urban
elements (buildings, locations, landmarks, etc), are linguistic signs, they
constitute a vocabulary, organized by a syntax made up of spatial relations
into a language. Kevin Lynch (1960) initiated the analysis of such a
language and its imperfections in Los Angeles, and his surveys, confirmed
by Banham's (1971) analysis, indicate the coexistence of two different
levels of language separated by a void.

At the neighborhood scale, we have a local language: the buildings, the
pedestrian paths, the street situations, and their happenings are known in
considerable detail to those who live nearby. Neighborhoods exist where
such concepts as orientation, contiguity, and proximity are correctly
perceived, understood, and transmitted. This sort of local language is a
static one, based on the relative position of buildings and alignments. It
is also a *spoken* language, as it enters into conversations between
neighborhoods, and a *concrete* language, appealing to the visual, hearing,
and olfactory senses as they are combined in everyday experience. In
Los Angeles, such neighborhoods seem to be less extensive than in
European cities, probably because pedestrian movements, the best way of
discovering them, are so much more limited. Lynch, in a further series of
studies in Los Angeles, has shown how the sense of familiar neighborhood
enlarges with an increase in the social level of the person surveyed. In
Los Angeles, as elsewhere, poverty means a shrinking of the perceived and
known space.

On the other hand, we also have a regional language, a language suitable
for describing things at the metropolitan scale built on the wide freeway
network of the city. This is both a *kinetic* language (made up primarily
of impressions of speed, going up and down, turning, stopping abruptly,
and so on), as well as an *abstract* language, because the car driver must be
totally obedient to traffic signs, giving up any confidence he may have in
his own sense of direction and distance. When you drive in Los Angeles,
it is practically impossible to perceive and master the orientation of the
concrete ramps as they leave the main freeways and are intertwined in
complicated three-dimensional figures. A person driving along Hollywood
Freeway to the east trying to reach the Civic Center, will see the famous
tower topped with a tetrahedron appearing close on his righthand side.
He may find it very hard to believe signs ordering him to cross to the
extreme left lane, but if he does not obey blindly he will rush past the

exit and find himself several miles further on before he can turn back to his destination. Obedience to paradoxical indications cannot be thoughtful: at 65 miles an hour, on a crowded five-lane freeway, there is no time for thought. As a result, the spatial language is *written* on huge green signboards in common English—not even in symbols as in Europe— so that the abstract language denies and overcomes the testimony of the senses.

These two languages, composed of two systems of relations between signs located in space, are well known to Angelenos who 'speak' them more fluently than any driver in other big cities: "I learned to drive in order to read Los Angeles in the text" (Banham, 1971). But the problems stem from the lack of any link between the two languages, and it is here that the feeling of disorientation originates, a feeling of anxiety that bothers visitors and makes them hate the City of Dreams. Los Angeles cannot really be considered as an undifferentiated and formless urban mass, yet these are precisely the common adjectives and clichés which many people use to describe the city. There is a missing link, a missing *language*, between the small neighborhoods and the large freeway network. Both are quite familiar to most Angelenos, and even to visitors after a few days, but their relationship remains unclear for many people. As Lynch (1960, pages 40–41) noted:

"When asked to describe or symbolize the city as a whole, the subjects used certain standard words: 'spread-out', 'spacious', 'formless', 'without centers'. Los Angeles seemed to be hard to envision or conceptualize as a whole. An endless spread which may carry pleasant connotations of space around the buildings, or overtones of weariness and disorientation, was the common image. Said one subject: 'It is as if you were going somewhere for a long time, and when you got there, you discovered there was nothing there, after all'."

A subtle remark by Banham (1971, page 213) confirms the point, when he noted "coming off the freeway is coming in from outdoors". He observed that as many cars left the freeway for the residential streets the women passengers would automatically check their makeup in the mirror on the back of the sunshades, implying that leaving the freeway actually symbolized the entrance into a friend's house. It almost seemed that there was no intermediary space between the freeway and the door of a residence.

One means people use to read the urban structure is to replace the missing spatial language by a temporal one; for example, the differences in the ages between districts orient themselves in the urban space (Lynch, 1960, page 41):

"The apparatus of regional orientation included ... a central gradient of age over the whole metropolis, evidenced in the condition, style and type of structures appropriate to each era in the successive rings of growth."

Banham also commented upon the importance Angelenos place on the signs of the past. These have partly a symbolic value for people, but they also help to identify landmarks (such as the famous Olvera Street), and to organize the urban structure through a useful chronological classification. The very speed of change in Los Angeles disorients inhabitants and visitors—"Each time I come, I don't recognize things anymore, everything has changed"—but at the same time the changes are used by them to introduce some order into the immense city. It is almost as though Time were projected onto the spatial plane, as if architectural signs, expressing age through their ordered distribution through space, took on a particular meaning. There is no question that a temporal language, although still imperfect, appears in order to organize the urban landscape.

One feels a need for a dictionary of urban signs, and for a corresponding grammar to describe their syntactical relations. Research in this field has only just begun, and it is impossible to go beyond a rough statement of the question at this point. Nevertheless, the multitude of urban signs in Los Angeles, so diverse and yet so aggressive, may be classified under three main headings of signification: Nature, Culture, and Power. These themes can be related in pairs, but we can also summarize the combinations in a more convenient way (table 3), at the same time making it clear that 'Domination', for example, which signifies Power, is also related to 'Control Over Nature', and so on. The categories are not necessarily exclusive.

Inside each principal Signification, the Signifieds are opposed dialectically: Nature is signified by various signs of openness to the external world, which receives it, in a sense, passively. But Nature is also signified by signs of control (over light, temperature, space, etc) which are distinctly active. In the same vein, Culture is simultaneously signified by a brutal and often flashy affirmation of modernism, but also by a passionate, moving, and sometimes even ridiculous search for legitimacy by creating today links that signify the past.

The same dialectic operates for Power. Pure aloofness and isolation would actually undermine Power, so one must retire from the crowd, and yet at the same time locate oneself in relation to the crowd in order to express domination over it. Nothing signifies Power better than those Los Angeles houses roosting on a hilltop, a summit location which can only be reached after a long drive over a virtually deserted road, winding for miles inside a canyon. Driving up Beverly Glen or Laurel Canyon one feels far away from any city, as though one were in the middle of a wild and foresaken mountain area. Eventually one reaches a huge villa standing on the crest, a dwelling that deliberately turns its back on the wilderness, built entirely in order to offer a fantastic panorama over the city. The owner has spent enormous sums of money to separate himself and his home from the city, only in order to come in touch with it again from a superior or dominating position. Even on the ocean shore, certain houses tend to dominate the beach and the sea through cantilevered rooms, with wide windows jutting out like observatories. These sorts of sites, which speak so clearly of Power, are very scarce, and are reserved accordingly for only the wealthiest. However, it is always possible to mulitply them artificially by placing them on the tops of high buildings, and in this way the penthouses of Los Angeles are the dominating eagle nests of the not-quite-so-rich families.

Similar nuances, containing within them the same dialectical seesaws, appear in the language of Isolation. The point here is to show how different one is from one's neighbor. The extremely rich use sheer space: they build their homes in grounds so large and so sheltered from others that one can hardly catch a glimpse of the houses themselves. Since respect and deference are ensured beforehand by the neighborhood's fame, the owner can do without the admiring contemplation of the casual passerby: in Bel Air, the most expensive part of Beverly Hills, the sidewalks are missing so that one cannot stroll by, but only drive through. Stopping is forbidden.

Table 3. The urban language of Los Angeles.

Sign	Signifiant	Signified	Signification
Open views			
Hilltops ... penthouses			
Bow-windows ... glass walls	Space		
Patios ... verandas, balconies			
Drift toward the beach		Openness	
Basins, fountains ... individual swimming pools	Water		
Parks ... parkways ...			
Freeways with plastic shrubs	Green		Nature
Gardens ... lawns ... plants inside the home	space		
Electrical light			
Air conditioning	Artificial	Control	
Access control	power		
Wood ... stucco ... concrete ... plastics ... light ... steel	Artificial materials	Modernism	
Pseudostyles:			Culture history
Antique, Colonial, Alpine, Spanish revival ...	Past	Legitimacy	
Decoration: neocolonial, pseudomedieval			
Barriers			
Hedgerows, wide lawns, parks ...	Isolation	Separation	
Housing			Power
On hilltops			
On the top of buildings	Elevation	Domination	
On the shore, above the ocean			

To 'speak' in this way, however, is far too expensive for most people. When a parcel of land is smaller, the relation swings dialectically: the point is no longer to separate by concealing, but to separate by showing off—as if the owners were afraid the passerby would not be impressed enough by their houses. They feel obliged to attract a person's attention, while at the same time keeping him at arm's length. This is the function of the grass lawn in front of every true American house, at least when it is occupied by a middle-class family. The lawn is carefully groomed, so denoting its important role and the way it is intended to impress the wayfarer, but it also has the function of forbidding him to come too close—"Keep off the grass!". In most cases, a narrow concrete path or driveway limits one boundary of the private property, and so shows the visitor where he should stand.

But how does one create a distinction from the mass when one has only limited means and is part of the mass itself? There is no question anymore of delicate refinements, quiet understatements, or huge manorial grounds; there is not even the opportunity to plant a limited but haughty piece of lawn. Instead one must express separation more crudely, and put all the more emphasis on the property that is itself so small. In danger of being lost in the crowd, a person using such social 'talk' cannot afford to murmur subtly—he must scream. Low fences (so low one could easily stride over them) appear around gardens the size of a handkerchief, symbolic barriers carrying out their unique role of saying "This is not yours, *I* own it!". Fences in the poorer districts bear the same relation to the wide lawns and parks as slang bears to correct and elegant language: they represent a popular way of 'talking', a way more direct, with more immediate images and allusions, with richer metaphors and allegories—in fact, a more rhetorical way of speaking. Fences are often painted or decorated in the most various and imaginative sorts of ways. Narrow gardens are embellished with ceramic statues; the small plots are dotted with earthenware cats, artificial ducks or storks, and even Snow White's dwarves. Such slang can be quite subtle: the dwarves try to be nice to the wayfarer and smile at him, even as they guard the property and keep him away. There is also a confused attempt to recover a sense of legitimacy on which to base a newly acquired property, and this effort to root oneself in the past is typical of such a modern city. Everybody knows that in medieval Europe dwarves ran through the land!

The search for legitimacy, the need for cultural landmarks endowed with historical meaning, are particularly noticeable in Los Angeles. Even if we make allowances for some of the fantasies characteristic of American society, it is still true that there are few cities in the United States where people try so hard to signify the past. The need and expression of a past are not unrelated to income, at least for those who have recently acquired some money. If poor houses try to relate to the Conquista with their pseudo-Spanish-colonial style, some of the bigger and richer ones go straight back to the Middle Ages, featuring towers with battlements, portcullis chains, flickering torches (burning gas), and armor in the entranceway that can be glimpsed through the pseudo-Gothic porch. Some houses even go back to Antiquity, with frontages based on Greek temples, and columns decorated in a style which could be said to be 'composite'—in the complete and literal sense of the term.

At the same time, resorting to the past in this way is negated by the pride of modernism. In contrast to most big cities on the East Coast, Los Angeles seems to be particularly proud of using the most modern construction materials with great freedom and imagination. The city evolved very quickly from wood to stucco to concrete and steel; it then moved on to plastic, and today to artificial light in order to heighten and accentuate its modern architectural forms, a visual characteristic which constitutes an outstanding feature of the city. Los Angeles is lit by floodlights, searchlights, fluorescent gas tubes with variegated colors (neon, xenon, argon), gas torches, as well as light in constant movement— flickering flashes, changing luminous signboards, and rotating restaurants lit like old-fashioned lighthouses. These decorative materials, extremely fluid and plastic, have led to fantastic expressions in architecture, where language is changing into myth, and sign into symbol.

It would be too easy to make fun of a language that is sometimes so excessive that it falls into unintentional *kitsch*, but such smirking and fingerpointing would be unfair. In the ghettos, flashy decoration is a way of expressing human personalities limited and crushed by too many taboos, and elsewhere the exuberant language expresses a satisfaction which may appear illusory, but nevertheless forms a basic component of the American way of life. The decorative language denotes a mixture of audacity and candidness which some Europeans tend to brand as uneducated simplicity, but it actually contrasts very favorably with the pretentious vulgarity and the complete lack of personality so typical of European suburbs.

Given the lack of previous analyses along these lines, it is difficult to guess how this sort of language will evolve in the future. We can, however, look for clues in that part of the language expressing certain aspects of nature, because this is the area where changes have been very marked over the past thirty years. For example, the ability to generate a sense of openness into external space, while at the same time creating an expression of domination, has been increasingly constricted to heights that have been created artificially—essentially at the tops of apartment towers. This may well be one of the reasons why they have developed so rapidly throughout the city, since today the mountain slopes have been totally occupied and are greatly overpriced.

There has also been another trend in the language, as nature has been attacked from all sides and tended to vanish. But evolution went on, from the parks and tree-lined avenues where people once rode horses, to the first Pasadena Parkway in 1938 lined with groves of trees and shrubberies, to the freeways of the 1960s still divided by natural bushes. Since the shrubs died from pollution, an extraordinary project tried to replace them in 1970 with plastic ones, 'bushes' that were 'ever green', smoke-resistant, washable, and unaffected by smog. Words in the urban vocabulary were thus transformed in an interesting effort to express the same signification, namely, nature. Individual parcels of land have followed a similar linguistic evolution: lawns try to say today what tree-planted gardens of old expressed; and for those living in apartments, and not wealthy enough to grow plants on a terrace, there always remains indoor gardening, which has now grown into a large industry, particularly among young people, representing the more recent word form that tries to say 'nature'.

Symbolic value and urban structure in Los Angeles
There is no question that Los Angeles is one of the great cities of the world where urban symbols, and their unconscious expressions in myth, are present to an overwhelming degree. Unfortunately, this does not mean that they are always clear and distinct, nor are they always easy to analyze. Any American city is characterized by a number of basic symbols, and these can obviously all be found in Los Angeles too. The Central Business District often expresses its economic power through the phallic symbol of the skyscraper, although this particular value appears less prominent, and even quite subdued, in Los Angeles. In comparison to the overwhelming Wall Street towers of New York, or the elegant Hancock Building and Sears Tower (the highest in the world) of Chicago, the center of Los Angeles displays an unimpressive, second class, almost shabby appearance. Except for a beautiful new complex made of cylindrical glass towers, its skyscrapers are too few, too banal, and too widely spread across the city to make a worthwhile and potent impression.

What is no less surprising is that the classical myths of the West Coast are also poorly represented—the Myth of the Far West and the Myth of the Pacific. Several restaurants try to look as if they were part of the decor in a cowboy movie, but the best saloons of the Old West, and the truest evocations of the Forty Niners are to be found in San Francisco. In those earlier days, Los Angeles was still a small town and not really part of the classical West. Similarly, it contains quite a large population of oriental origin today, but its Chinatown and Little Tokyo are not particularly visible in the huge urban area, and they are no match for the Chinatown of San Francisco.

Nevertheless, if the standard symbols and myths of American cities are not especially prominent in Los Angeles, the city has generated several new ones which typify it exactly. One of the most prominent is the *Star Myth*, a symbol of success that was born in Hollywood and rooted in Beverly Hills. It is related to the *Gold Symbol*: not gold as a source of power and fear, not the slightly mysterious and occult symbol as in Wall Street, both concrete and abstract at the same time, but rather gold which shines and sparkles, which wants to exhibit itself, and derives its success from its very exhibition. Baudrillard quite rightly underlines the ambivalence of symbolic values: since they depend on Freudian subconscious mechanisms, they do not respect the logical principle of noncontradiction, but attract and repel simultaneously. Freud has shown that gold is unconsciously perceived as the equivalent of faeces, the infantile gift of the child to his mother.

In Los Angeles, show business is symbolically related to gold in its natural flake or nugget form, and these words are present in the signs of many bars and restaurants—just as they are in Las Vegas. This glittering aspect of glamor is dialectically completed by its opposite, the aura of 'corruption' which Hollywood tries so hard to deserve, as gold and faeces are mixed in Freudian images. A whole mythology of vice and perversion, of wallowing and unmanning luxury, of success leading to disaster, of alcohol (and more recently drugs) ruining a promising career, keeps flourishing in cheap and widely-read newspapers, and even in the B films made in Hollywood itself. The city creates and thrives on stories of its own (supposed) debasement. Such myths are much too complex to take them apart completely and detail all their inner mechanisms here. In fact they are frequently made up from more elementary myths: Cinderella ascending to glory; Gyges spoiled by the gods, trying to win them over by offering them his ring (that is, making a spectacular gift to the Salvation Army), to find his offer disquietingly refused (God is not for sale at such a low price); the myth of the Capitol so close to the Tarpeian Cliffs; the myth of the Vamp, the venomous woman, so attractive and so destructive, a modern siren calling from a bar stool (again the intertwined connotation of gold and faeces); the myth of Balzac's 'Peau de Chagrin', with the image of life's candle burning at both ends, of lives running to self-destruction like the tapes telling the story in 'Mission Impossible'. All the ingredients are here to excite the dreams of the masses: they get the intoxicating impression that success is around the corner, just within the reach of their outstretched hands; but at the same time they are reminded that the social hierarchy is stable after all, and that in a really 'happy ending' the unheard-of-success that upsets social rules is finally and inevitably broken to pieces. One must keep to his position and his dreams.

All these myths are epitomized in a remarkable way in *Hollywood-Babylon* (Anger, 1976), a long list of Hollywoodian glories, excesses, and disasters: 'Fatty' Arbuckle, among the leading comics of his time, accused of murder and ruined; Buster Keaton, a genius dying in misery, broken by his son's leukemia; Erroll Flynn, the gifted swashbuckler under a conviction of statutory rape which almost ruined his career; and then all the lives, several terminated by suicide, that were broken in a

few months when sound movies replaced the silent screen. Perhaps the
most poignant symbol is the actress, unemployed after playing in thirteen
films, who drove one night up the hill to the gigantic illuminated letters
reading HOLLYWOODLAND, undressed, and jumped naked to her death
in a cactus bush from the last D, the thirteenth letter. Here is success,
failure, destruction, movement, light, superstition, sex, linguistic sign,
hope, and hopelessness—the symbol is complete.

Nathaniel West (1939) has described in his novel the violence, the
passion, and the bitterness of Hollywood life. In fact, limits between
symbols and reality vanish: if Hollywood portrays the world through
books, films, and shows, it also turns dialectically upon itself and so
becomes its own object. It is the story of the photographer photographed,
as reality becomes so totally symbolic that symbol becomes reality.
From Nathaniel West to Andy Warhol, how many novels and films do we
have about novel writing and film making, about writers, producers, and
actors? The narcissism of the actors and actresses contaminates the
whole system: it is both a mark of its success, as well as one cause of its
failure. And, quite typically, the cinema virtually left Hollywood in the
1960s to rediscover the world.

The hopeless mixture between reality and fiction generates the *Myth of
the Fantastic*, exemplified in the architecture over the whole city, but
particularly prominent in the show business districts themselves.
Grauman's Chinese Theatre, the "ultimate shrine of all the fantasy that
was Hollywood" (Banham, 1971, page 113), is loaded with redundant
symbols: prints of people's feet and hands are actually stamped in the
cement in front of this famous movie theatre like non-written and
indelible signatures. On the sidewalks along Hollywood Boulevard, bronze
stars embedded in the cement feature the names of famous personalities
from the radio, movie, and television industries, and people walk on them
as if they were entering a cathedral filled with the tombs of vanished
royalties. Fantastic buildings realize a succession of dreams in plaster and
metal: Simon Rodia's fantastic towers; the house, built in 1925 for
Henry Oliver, looking like the eerie scenery inspired from Hansel and
Gretel; and, perhaps more than anything else, Disneyland, the first
example in the world of a real city coming straight from fairytales.

Banham has described the fantastic styles used in restaurants.
Metonymy is the rule here, and the part speaks for the whole: a plastic
totem and three illuminated arrows announce the 'Tahitian Village'; a
huge pineapple, five meters high and flashing green and yellow says
'Hawaiian Food'. The symbols form such a luxuriant jungle that even
epiphytes grow in its shadow, parasitic symbols grafted onto others.
The old oil derrick kept in Santa Monica as a symbol of oil wealth, is
now dressed with painted wood representing a medieval embattled tower,
which is dressed symbolically in turn as an Austrian beer parlor!

The whole city looks like an immense stage, where cardboard pieces of scenery are dispersed in the most fantastic ways. An old photograph shows Universal Studios during the 1930s while they were filming *Ali Baba and the Forty Thieves*; the open air studios are so vast, and their pseudoexotic sceneries of a totally false Middle East blend so well into the surrounding townscape, that it is difficult to know where fiction stops and reality starts. In Los Angeles one never knows for sure.

Baroque forms, *kitsch*, romanticism, and fantasy expressing audaciously symbolic values, are the last thing that the European traveler expects to find. These are not the things that characterize the United States as they are usually depicted, with efficient and cold robots led by computers with a touch of skeptical English humor. On the contrary, few visitors expect the overornate style from eastern Europe, laced with German 'gemütlich-keit', and set in scenes built straight from Grimm's fairytales. The true originality of Los Angeles, among all other American cities, lies not so much in the important role of symbolic value (after all, this is significant in all cities), but rather the freedom of its expression. It is, in fact, the symbol of its libertarian anarchism.

The *Myth of Libertarian Anarchism* clearly contrasts Los Angeles with the large metropolises of the East Coast. At the far western end of the continent, people tried to realize a Utopia by turning their backs on the urban failures of the East and Midwest, a Utopia that is expressed by a total freedom for everyone to express themselves in their own symbols and to live by them. This myth also explains the vogue of the automobile: it is, of course, a means of transportation in a dispersed city, but it also forms a distinct means of expression. Nowhere in the United States, and perhaps nowhere in the world, can one observe a larger variety of forms and colors in the bodies of the cars. Many have been repainted, redecorated, cut, welded, and reshaped into the strangest and most personal forms. Los Angeles is the birthplace of the 'dune buggies', small open cars with plastic bodies sparkling like electric guitars, made to drive on dunes and beaches. They are the humble Volkswagen— the People's Car—dressed up in finery to show off. The same aggressive freedom appears in the incredible motor bikes, with the front wheel pushed far ahead of the engine, and the driver lying lazily back as if in bed, and in the uniforms of the Hell's Angels mixing hippie parapharnelia with the helmet and Iron Cross of the Wehrmacht. This is also the home of the bright green Rolls-Royce upholstered with pink fur. In all these forms of myth, the point is not so much to show off one's money, but to bang one's fist on the table, and to state noisily and freely one's idiosyncrasy. Anarchism passes over into another symbol—movement.

The Myth of Movement is based on the importance of change in such a volatile city. It is not simply rooted in the everyday experience of rapid urban transformations; in Los Angeles, movement itself becomes a myth.

The inhabitants tend to exaggerate its importance, because it is not only a fact, but also the very symbol of a Utopia being realized everyday. Of course, it is true that many illusions and hopes have crumbled, particularly since the 1970s when the local economy suffered a severe setback, the population stopped growing, and change itself slowed down as the city went into a period of painful self-questioning. Nevertheless, the myth of movement is still there, and quite as important as the sense of movement through time is the sense of movement over space. It is here that the main role of the automobile appears—as a symbol of individual freedom. What Banham calls 'autotopia' is obviously a myth: driving a car at high speed on a freeway (and notice the connotation of the prefix 'free') should symbolize liberty, but it actually submits the driver to a severe discipline and a total obedience to traffic indications he does not understand. Myth, as always, is deceitful, and ideology lurks behind it.

Conclusion

In Los Angeles, as elsewhere, all the different values combine to constitute the satisfaction that people can experience from their homes and their neighborhoods. In a market dominated by competition, they may all be integrated in the concept of land price or rent, but competition is never perfect, and it is often greatly distorted by publicity. Racial and social prejudices, for example, build almost opaque barriers, so that correspondences between price and value are likely to be weak. Even more important is the fact that the most interesting values (use, sign, and symbolic values) are not comparable; each depends on a different qualitative judgement. They are likely to be evaluated quite differently from one social group to another, from one age group to another, as well as between different cultures and different educational levels. It will probably remain impossible for quite some time to measure them precisely.

Driving through Los Angeles, one can easily see all values combined, although in a rather confused way: hence the first impression of a disjointed city, a city with a lack of structure—an impression that is actually misleading. Yet everything seems to contribute to building the wrong impression: the dispersion of the business centers; the immensity of the agglomeration; the variety of perceived land values which cannot be expressed in a common unit of exchange; and the superimposition of two spatial languages (local and regional) without intermediate links connecting them—the surface streets and the freeways. A mathematician once said that Los Angeles was a discrete topological space, made up of equidistant points: any two points chosen at random are almost a constant distance apart in driving time, since it is time-consuming to get on and off the freeways, although driving on them is extremely fast (Ulam, 1962, page 174). This tongue-in-cheek mathematical description, that all points in the city seem almost the same distance from each other, expresses admirably the apparently amorphous nature of the city's urban structure, and its excellent overall accessibility by automobile.

Yet, far from lacking in order, this immense urban structure of Los Angeles is an authentic language, revealing the true nature of the American city through its signs and its contradictions (Horkheimer and Adorno, 1972, pages 120–121):

"Even now, the older houses just outside the concrete city center look like slums, and the new bungalows on the outskirts are at one with the flimsy structures of world fairs in their praise of technical progress and their built-in demand to be discarded after a short while like empty food cans. Yet the city housing projects designed to perpetuate the individual as a supposedly independent unit in a small hygenic dwelling make him all the more subservient to his adversary—the absolute power of capitalism. Because the inhabitants, as producers and as consumers, are drawn into the center in search of work and pleasure, all the living units crystalize into well-organized complexes."

References

Anger K, 1976 *Hollywood–Babylon* (Dell, New York)
Bachelard G, 1957 *La Poétique de l'Espace* (Presses Universitaires de France, Paris)
Banham R, 1971 *Los Angeles: The Architecture of Four Ecologies* (Penguin Books, Harmondsworth, Middx)
Barthes R, 1967 *Systèmes de la Mode* (Collection Tel Quel, Paris)
Baudrillard J, 1968 *Le Système des Objets* (Denoël, Paris)
Baudrillard J, 1972 *Pour une Critique de l'Economie Politique du Signe* (Gallimard, Paris)
Clark W, Cadwallader M, 1973 "Residential preferences: an alternate view of intra-urban space" *Environment and Planning* 5 693–703
Horkheimer M, Adorno T, 1972 *Dialectics of Enlightenment* (Seabury Press, New York)
Lynch K, 1960 *The Image of the City* (MIT Press, Cambridge, Mass)
Perloff H, and others, 1973 "Prototype state of the region: report for Los Angeles County" School of Architecture and Urban Planning, UCLA, Los Angeles
Ulam S M, 1962 *Adventures of a Mathematician* (Charles Scribner's Sons, New York)
West N, 1939 *The Day of the Locust* (Penguin Books, Harmondsworth, Middx)

Geography and the realm of passages

Erik Wallin

The subject matter of geography

The subject matter of geography is peculiar in many respects. Whereas other disciplines have restricted and well-defined domains of subject matter, geography seems to be able to handle *anything*, at least if 'anything' can be given some spatial attributes, or if 'anything' can be attributed some spatial consequences. Yet I think it is possible to qualify these 'anythings' in such a way that geography emerges as a discipline with a most interesting subject matter of its own. In my own perception of geography, most of these 'anythings' can be seen as aspects of different *no-things*, entities that are no things, but nevertheless are ingredients of the stuffing of the world. In what follows, I will call these entities *passages*, and the stuff they constitute will be called *communicative intermedia*.

Passages divide and unite the world. They work as intermediates and as mediators between different parts and sides of the world. In fact, most 'parts' and 'sides' that appear in the world are *caused* by such passages. Like doors they both separate and integrate 'indoor' and 'outdoor' phenomena. To pass such doors is often to engage in a *communication* in the literal sense of entering a communion or sharing a common 'inter-esse'. In order to avoid trespassing, and to secure indoor treasures, many rules, norms, and rites are associated with such passages. I will discuss these things later, but let me first outline how geography as a scientific discipline can be related to the realm of passages.

Geography is almost unique in that it still has one foot in science (as physical geography), and one foot in the humanities (as human geography). The problems that emerge when trying to integrate these two approaches in the study of social life are essential for geography as a scientific discipline. In fact, the main subject matter of geography can be located to *the societal passage*—the passage where nature is being cultivated and domesticated and where culture is being naturalized, taken-for-granted, and (often) alienated. The study of the development of Virginia both as a landscape and as a 'mindscape' by Jean Gottman (Gottman, 1955) is a case in point.

Another aspect of geography is the concern for general and overall constructions, both when making *models* of the world (descriptions) and when making *plans* for the world (prescriptions). Geography is, in contrast to most other disciplines, *synthetic*. It is not only a question of making a synthesis of the world *as it is* (the reality), but also a question of making a synthesis of the world *as it ought to be* (according to the imagination). And, to make things still worse, it is also a question of making a synthesis of *these two versions* of the world! Consequently the geographer has to take part in the battle between realism and idealism, a battle that is taking place in *the realizational passage*—the passage where reality forcefully presents itself, and where dreams seek to come through.

Prevailing geographical methods, like cartography and planography (the art of making plans), are highly biased towards territorial space and to phenomena that, quite literally, *take place*. As a consequence, phenomena that *take time* and develop in time have been neglected. A traditional map of a town, and a traditional town plan, are cases in point. Time-geography, and the attempts to create a *time–geo-graphical* language, can be seen as one of many efforts to overcome the shortcomings of traditional geo-graphical languages, both when making models and when making plans; for instance, through time-use-orientation. But not even a time–geo-graphical language will do. What is needed is some sort of context-dependent, 'bio-graphical' language that can deal with human actions, biographies, and phenomena that, in addition to taking place and time, also *take meaning*. After all, life on earth is not meaningless.

The concern for meaning makes it necessary to look more closely at *the passage of everyday life*—the passage we all must pass every day in order to be present. Depending on what kind of language we are operating in, and depending on what attitudes we adopt, different snatches of life will appear when we try to grasp what is going on in our day-to-day world. It is important, therefore, to consider the expressions of everyday life from different perspectives, and interpret them through different languages. The art of shifting scales, both in time and space, is inherent in the geographical way of thinking. This art could be extended further, making it possible also to make *shifts of languages*. We must, so to say, learn to talk the language that the phenomena speak. Geography can, I think, make important contributions to such an endeavor, to shift scales not only in space and time but also *in essence*. The present essay is such a contribution.

The nature of passages

Passages must be considered as entities with their own ontological and epistemological nature. I have deliberately chosen the word 'passage' to represent them, because the meaning of this word is sufficiently rich, ambiguous, and context-dependent. A passage is a passage *in all senses* of the word—an act of going across, a corridor, a voyage, a combat, a short extract from a speech, etcetera. The kernel property of a passage is its twofold capacity *to distinguish and to connect* different phenomena *from and to* each other. In fact, the possibility of identifying differences at all is *due to* corresponding passages. We have no really effective language to deal with passages, perhaps because they are already inherent *in* our language to such an extraordinarily high degree. Consequently, they are not very easy to extract, grasp, and talk *about*. Passages are (for the same

principal reason that a fish can hardly conceive the water in which it lives)
hardly conceivable.

Passages work as mediators and coordinators between the world,
considered as a *host*, and the phenomena, considered as *guests*. *The
passage of the present*—the passage where everything occurs—must be
looked upon as a series of *exchanges* and *rounds* in which the 'presents'
of the host are reaped and shared among the guests. Both the host and
the guests can appear in different guises. The three modes of revelation
of the host that I will single out for discussion are the host as a donor of
place (space), as a donor of time (life), and as a donor of meaning
(language). The corresponding modes of appearance of the guests are
physical bodies that take place, living beings that take time, and social
persons that take meaning. The corresponding 'rules of the game' are the
laws of nature, the principles of life, and the rules of language.

In daily life these different 'games' or modes of exchanges are, of
course, mixed together. The spatial and territorial movements are closely
connected with corresponding social and linguistic acts; for instance, when
passing a door. In fact, the 'physical' crossing of a 'real' threshold can be
seen as the prototype for more 'mental' or 'artificial' passages. The
different *rites of passage* that Arnold van Gennep has identified in many
cultures can be regarded from such a perspective. He writes (van Gennep,
1960, page 20):

> "Precisely: the door is the boundary between the foreign and domestic
> worlds in the case of an ordinary dwelling, between the profane and the
> sacred worlds in the case of a temple. Therefore to cross a threshold is
> to unite oneself with a new world. It is thus an important act in
> marriage, adoption, ordination, and funeral ceremonies."

The passages of everyday life constitute a mixture of different *communicative
intermedia*. These channel, modulate, and regulate the behavior of the
'passengers'—those who are communicating in the actual medium.

Space, life, and language as communicative intermedia

Space, in all its shapes, forms, and usages, is of course the most primitive
communication intermedium. Space is the prototype for *a consistent
whole*. Space both permits and compels the most primitive forms of
communication: physical contacts and penetrations. The passages of space
appear as paths, routes, lines, and hollows in the gravitational and other
fields that support space. In a sense, space is more *in*-sistent than *ex*-
(s)istent, as it forces matter to follow these passages, and restricts the
exchanges and interactions in certain ways, according to the laws of nature.

Considered as a host, space gives room for phenomena to take place in.
To exist is to *ex-(s)ist* in the sense of standing (out), of enduring, all the
burdens, pressures, and demands from space. One such pressure is brought
about by the *competition* for space among the guests in space. Formations
that do exist have literally *passed* the general passage of time. They have

not only passed, but have done so in a *suitable* way. They have been
dressed in shells, membranes, and other virtual costumes that give adequate
resistance to primitive forms of communication, such as the bombardment
of elementary particles. In general, one could recognize different forms of
im-munication—the establishment of barriers and shelters against 'outdoor'
disturbances. Through such im-munication, formations are created that are
immune against certain demands and pressures from the local space in
which they exist.

Most of the transports and transformations that occur in space are
irreversible. There are, so to say, no return-tickets available for the
passengers, and, because of the general passage of time, the previous order
is more or less impossible to restore. Some of the passages can, in fact, be
considered as *local entropy minimizers* that bind up local free energy into
complexes of higher order. They force the passengers, in the form of
atoms and cells, to become members of different *communities*—organic
wholes with some shared common interest. Life can be considered as one
such common interest.

Considered as a communicative intermedium, life is the prototype for *an
organic whole*. Life both permits and compels some of the most basic
modes of communication: struggle (for life) and sexual intercourse. The
passages of life take the form of semipermeable membrances, orifices,
hides, and other *intermedia* that, on the one hand, differentiate the
interior from the exterior, and, on the other hand, integrate them with
each other. Life is, in a sense, more *per*-sistent than ex-(s)istent, as it
continues to develop in new and other forms, irrespective of what the
individual creatures are doing. Living beings are forced to follow the
principles of life, including the principle of individual death.

Considered as a host, life gives a space of time to the guests to live
through. But life, as a host, seems to take more care of the specific types
and models of matter (the species) than the individual copies or specimen
(the individuals). In fact, life *itself* can be considered as the communicative
intermedium that makes *re-presentation* possible at all, *both* in the literal
sense that formations can be replicated and reproduced, *and* in the
ordinary sense, that formations can be translated and coded into models
and images that only take into account some essential features of the
'original' formation. René Thom has found that there seems to be a
subsystem in the metabolism of living organisms that simulates the metrical
configuration of the environment. He concludes (Thom, 1975, page 222):

"We might almost say, anthropomorphically speaking, *that life itself is
the consciousness of space and time*; all living beings have a common
representation of space, and competition for space is one of the
primitive forms of biological interaction."

It is important to note that living creatures do *not* react upon the
environment 'itself', but upon *the internal representation* of it. These
internal representations are certainly specific for each species, and depend

on the receptors that the organism uses to conceive what is going on in the environment. In the case of human beings, the internal representation is mainly done unconsciously. It seems more accurate to say that it is done by *our selves*, that is, by our specific nature, than to say that it is done by ourselves.

The im-munication that life provides is much more advanced than the corresponding im-munication in space. In fact, life makes it possible to *learn from experience*, as when children become immune from certain pathogenic agents after the first battle with them. It seems plausible that some of the linguistic stuff is treated in the same way, creating some immunity against subversive phenomena through some sort of defence mechanism. Otherwise the person would be broken down when confronted with 'reality'. Some of the passages of life work as virtual *developers* and *envelopes* that, on the one hand, provoke latent figures and bring forward new details of a crude pattern, and, on the other hand, cover such messages and stop the generation of new details. The translation and storing of the genetic code is a case in point.

Life seems to be eager to find many ways to store and transmit vital information and not only through the genes. Ordinary language can, in fact, be considered an *extra-genetic mode of transmission* of vital information. Waddington has elaborated this point of view further. He writes (Waddington, 1975, page 272):
"Evolutionary change involves the gradual modification of the store of genetically transmitted information. A few animals can pass on a meager amount of information to their offspring by other methods; for instance, in mammals some virus-like agents which have effects very like hereditary factors may pass through the milk; in some birds the adults may serve as models whose song is imitated by the youngsters, and so on. Man, alone among animals, has developed this extra-genetic mode of transmission to a state where it rivals and indeed exceeds the genetic mode in importance. Man acquired the ability to fly not by any noteworthy change in the store of genes available to the species, but by the transmission of information through the cumulative mechanism of social teaching and learning."
The main part of this social teaching and learning occurs as a consequence of the *practice* of everyday life. Important parts of the cultural heritage are transferred and transmitted through the *rites* of passages that were discussed earlier. The transfer goes mainly from the old carriers (the adults) to the young carriers (the children). The rites also work as local entropy minimizers as they bind up local unrest (for instance, unemployed people) into complexes of higher order (such as grandes fêtes). Through such arrangements, the common order is made visable, agreeable, and confirmed.

Language is, of course, the prototype for *a communicative intermedium*. Language both permits and compels some of the most advanced forms of communication—thinking and discourse. The passages of language appear as relations, liaisons, and other joints that make *articulation* possible. Language is, in a sense, more *con*-sistent than ex-(s)istent, as it fills gaps in the passages, adds missing parts in the expressions, and improves the passages of language in different ways. Language does, in fact, *give* meaning to phenomena. On the other hand, this interference makes it necessary for the communicants to follow the rules of language; for example, the rule that 'normal' conditions should *not* be expressed as if they were unknown. For instance, the report that "today, the captain is sober" violates this rule if, usually, the captain is sober.

Considered as a host, language has many *spokesmen*. These spokesmen are often partial and make a strong separation between those who are included in the actual community and those who are excluded. If space makes a distinction possible between what *in-sists* (fields, forces, and so on), and what *ex-(s)ists* (particles, bodies, and so on), and if life makes it possible to distinguish between an *in-side* and an *out-side*, language makes it possible to discriminate between *insiders* and *outsiders*. On the other hand, this very discrimination stimulates a communication between "You-There" and "Our-Selves-Here" *because* of the difference.

The im-munication that language provides is of utmost importance for the channeling and distribution of the burdens of social life. The im-munication takes the form of real suits, costumes, uniforms, headgears, and virtual *hoods*—such as child-hoods and man-hoods, that filter and sort out what kind of demands, taxes, and 'courses' should be addressed to the different communicants. To a certain extent, ordinary occupations can be regarded as such 'immunities' and 'fortifications' against societal demands. Males, for example, have traditionally entered into occupations that are immune against the cries and demands from the babies and children in society. This is, I think, the basis of the sexual division of labor. Women, in contrast to males, have not (yet?) developed such an immunity.

Towards a new realm of passages

It seems to be possible to work out a more systematic theory of levels of communicative intermedia, as outlined above. Such a theory would certainly be very close to the theory of *time as conflict*, that Fraser (1978) has developed. In this, the 'umwelt'-concept, created by the German biologist Jakob von Uexküll, has been elaborated further into what Fraser calls *the extended Umwelt principle*, saying, in brief, that the world of each creature is determined by the potential functions of the totality of its effectors and receptors. The world of an animal, its *Umwelt*, is the combined world-as-perceived and world-as-acted-upon. Fraser recognizes six semiautonomous integrative levels of communicative inter-media, called *levels of time* or *temporal Umwelts*. The first two of these levels, called by Fraser the "a-temporal" and the "proto-temporal" levels, correspond to space in my terminology. The third and fourth levels, called the "eo-temporal" and the "bio-temporal" levels, correspond to

what I have called life. The last two levels, called the "noo-temporal" and
the "socio-temporal" levels, correspond to what I have called language.
According to the theory of time as conflict, the general passage of time,
that is, development, must be considered as a whole series of *time's rites
of passages* through different stages or levels of time.

As an example, let us consider the phenomenon of a mosquito bite.
A human being can be considered as a physical *body*, consisting of atoms
of spatial matter. As such, the human being obeys the laws of nature;
for instance, when exposed to radiation, or when hit by another body—
let us say a mosquito. The atoms of spatial matter are constituents of
higher complexes, such as organelles and cells. Hence the bodies can be
considered as living *beings*, consisting of different cell-communities of
living matter. As such, they work according to the principles of life; for
instance, when confronting each other in the struggle for life. Endowed
with linguistic capacities, the human being can also be considered as a
social *person*, that follows the rules of language; for instance, when
experiencing a pang *as* a mosquito bite. Now, a single *event* for the human
being *qua* person must be regarded as a whole *epoch* for the human being
qua cell-community. What is happening on one level must be translated,
developed, and evoked into the language of another level; for instance,
from a biochemical to a psychophysiological language, and vice versa.

Such translations between different levels constitute the time's rites of
passages, according to Fraser. Obviously the phenomena appear in
different guises, depending on the language used for the interpretation.
There is a need for different *level-specific languages* in order to understand
fully what is going on in the world.

I cannot elaborate further here on what this new realm of passages
looks like, although elsewhere I have made such an attempt (Wallin,
1980). Let me only point out that a genuinely *new* common ground for
geography is certainly going to have to be based on a new conception not
only of space but also of *time*. According to Fraser, our conception of
time is a false generalization of conditions in our own, very speci-al,
Umwelt. He writes (1978, page 68):

"Perhaps, while talking about physical cosmology, we made the mistake
of neglecting the cosmologist."

And further (1978, page 149):

"Cosmological common time, then, is Kantian in that it is an *a priori*
form of perception; but it is also Darwinian, for it is a mode of
perception which has evolved through time. This curious stance,
consistent with the extended *Umwelt* principle, suggests that what we
recognize as the evolution of the cosmos in time, from past, to present,
to future, is not 'there' to be discovered but is a representation
appropriate to the nootemporal *Umwelt* of the mind."

The Present, considered as a threshold between The Past and The Future,
is really an *escalator*, not only in the sense of a moving staircase but also

in the sense that the world expands and unfolds through the passage of the present.

My own perception of what is at stake is that we still miss the rudiments of what must be crucial for *any* social science: a theory of *the psychology of the mass*—a theory of *our* (unconscious) *selves* in the sense given earlier. Considering the fact that our basic conception of the world is done by our selves, that is, by our specific nature, and considering the fact that our artificial environment to a large extent is a manifestation of our selves, we must look further into the development of our selves and the rites of passage that our selves have been subjected to in order to pass *consciously* through the passage of the present. We must, so to say, *take care* of the presents that the world presents to us. The ambition of this paper has been to create a suitable passage for the development of geography as an important caretaker of the world and its presents.

References
Fraser J T, 1978 *Time as Conflict* (Birkhäuse, Basel)
Gennep A van, 1960 *The Rites of Passage* (Routledge and Kegan Paul, Henley-on-Thames, Oxon)
Gottman J, 1955 *Virginia at Mid-Century* (Holt, New York)
Thom R, 1975 *Structural Stability and Morphogenesis—An Outline of a General Theory of Models* (Benjamin, London)
Waddington C H, 1975 *The Evolution of an Evolutionist* (Edinburgh University Press, Edinburgh)
Wallin E, 1980 *Vardagslivets Generativa Grammatik* (Liber, Lund)

Epilogue: A ground for common search

Gunnar Olsson

As this Epilogue takes form, a dream is coming true. Thus I remember
how a year ago I began preparing myself for that small meeting which
now is manifested in the present volume. After a chilly period of northern
isolation I was again ready to experience a handful of friends, to pay yet
another visit to Penelope's weaving chamber. There was little to loose, for
the tales about the Villa Serbelloni had me convinced that even if the
colloquium became an intellectual disaster it would at least be remarkably
pleasant.

And yet: Protestant ethics demanded not only that I do my homework
but that it should contain a strain of self-punishment to make up for the
good life to come. Since I could not imagine anything duller than a
month of solid geography, that was what I embarked upon. Indeed I
consumed large potions of the accumulated production by myself and my
fellow symposiarchs. What I got in return was a fresh impression of
recent intellectual history as reflected in some of the discipline's trend-
setters. Such macro/microscopic readings are doubly illuminating, for
they illustrate both that the production of a text is closely associated
with the making of its author(ity) and that reading is more demanding
than writing.

Misled by the title of our joint venture, my reading started as a search
for common ground. What I discovered, though, was neither a well
delineated subject matter nor an agreed upon philosophical approach, but
a determination not to be sucked into the dark caves of the seventies.
What was revealed was not a common ground but a common search, not
the safe acceptance of immutable dogma but an irresistable yearning for a
knowledge impossible to name.

The readings nonetheless indicated that despite our diversities and
searching uncertainties there was a shared interest in the cultural limits
of thoughts, languages, and actions. Even more precizely, our blurred
focus appeared to be on the concept of limit itself, especially on the
categorial boundaries of outer and inner, fact and value, countable and
uncountable. Gradually the most important relations became those
between ourselves as concrete human beings on the one hand, and the
surrounding wor(l)ds as abstract entities on the other. In a sense we were
obsessed by the skins of our bodies, by the various penumbrae through
which one gains contact with the institutions of others.

What initially had been a dull and dutiful preparation for yet another
conference was consequently turned into a moving experience of some
individuals' narcissistic search for meaning, for coherence in their

contradictory worlds, for approaches to the paradoxical relations of life itself. In this pursuit we came close to the mythical, for the mythical can be defined as that receptacle of accumulated insight in which we experience again what we never experienced before, that nexus of deep understanding in which I can recognize myself in others and others in myself.

Some genuine questions emerged: Why are the fundamental categories of the social sciences delimited as they are? What is it in the operational-ization of these categories that makes them seem natural today and not a decade ago or a decade hence? Who teach us to confine ourselves within firm boundaries when every child acts out the necessity of transcending them? How do I learn more about limits going to their limits, of demarcation lines becoming thinner and thinner? What made us talk about categories in 1968 and about limits in 1980? Is our present concern with limits a sign that we have reached them or already passed them? Why is it less dangerous to ask what you can say about current mythology than what current mythology says about you? Which insights did Nietzsche reveal in his talk about the love and shame of contradiction? Is the modern interest in Nietzsche itself a reflection of a man stupendous enough to practice his theory unto death?

Timely questions a year ago. Now with the next stage already summarized, the coherence is so striking that it is hard to believe. Knowing that each author was encouraged to write whatever he or she considered most urgent, the closeness of the individual pieces is amazing. Rarely have I experienced so deeply how even the most devoted individuals are children of our own times and places. As the haze from the seventies lifts, there is a glimpse of a common ground before us.

The vast differences in style notwithstanding, all papers focus on the same fundamental relation. This is the relation between individual and society. At the same time there is a strong tendency not to argue for the hegemony of one of these categories over the other, but instead to enter into the very glue which separates and unites them. There is an unusual sense of urgency in the writings, almost certainly because they practice philosophies which have proven illuminating to the authors themselves. The totality becomes solid for it reflects two truths approaching each other, one from the inside, the other from the outside. But here as elsewhere reader and writer are involved in the dialectics of master and slave, for the question of who I am is a question of how I delimit myself in relation to the rest, of the extent to which I let others get under my skin, of the words with which I sacrifice the totality of my own truths in order to communicate its pieces with the world at large.

Master and slave prompt questions of power, an issue written into every page of the volume. This is best seen in the constant problematizing of the level of analysis, regardless of whether the categories are those of phenomenology and positivism, backcloth and traffic, path and project, use and exchange, citizen and State, signifier and signified, thing and relation, you and I. But even though the choice of levels is often disguised under the operational cloak of data and theory availability, it is in essence an issue of the profoundest importance. When several authors refuse to choose, this is therefore in fact an ethical stand aimed at defusing the explosive / of Power/Knowledge. As a consequence, the methodological heterodoxy must not be interpreted as a sign of lacking dedication but as an instance of the belief that every attitude is partial.

Another aspect of power leads into the mechanisms through which individuals come to reflect the norms of given social and historical contexts. What is alluded to, however, is not the type of study that focuses on powerful individuals and repressive institutions but rather detailed inquiries into those micropowers which are so deeply ingrained in our thoughts-and-actions that only the most sensitive can notice them. In the critique of these prisons within and around lie encounters with the taboo, themselves tabooed, with those relations which are so forbidden that their forbiddenness itself is forbidden. In the state mode of production characteristic of late capitalism, it is these socialization processes that penetrate every aspect of our lives, that tighten inter-dependencies and produce actions of counterfinality, that legitimate procedures through which good intentions are turned into opposite results, that produce novel illnesses. In this context, an occultist may detect a message in the fact that our symposium was held at a place once called Tragedia. But, on the other hand, if life is a joke, then our writings about it should be likewise; just imagine the tragicomedy of Groucho in fact knowing more about people than Karl!

Projecting out of Serbelloni, I see forebodings of a decade so intellectually exciting that I cannot wait for it to arrive but must go after it at once. Our writings clearly demonstrate a decreasing obedience to disciplinary pasts, a feverish urge to cross boundaries. There is a vigorous determination to forge new wholes out of old parts, to understand through oxymoronic words sensitive enough to rerender the absurdity of living in the interface between one production mode and another. Perhaps the late discipline was too enchanted by the sirens of utilitarian duty and too fearful of the aesthetic. Perhaps we may live longer by our theories if they were founded more in beauty and less in political expedience.

The moral is that there are many ways to learn, many concepts for capturing reality, many languages for setting it free. But all languages are made of words held together by the rhythmic silences, the blank spaces, between them. It is therefore that a word is not always a word but sometimes a subscript, a house, a dress, a handshake, a glass, a silence, a silent-glance-over-a-glass. This is why it is inevitable to swim on in the Serbelloni spirit. WHICH MEANS: not always to yield to the social pressures of being dumber than you are; not always to be dead serious;

not always to approach truth as if it were inherently sad; but instead: to rejoyce in the rich languages of experience itself; to realize that the powers of reality are more insulted by laughs than by tears.   Thus there is a key to both the individual Freud and the collective Jung in Molly's remembereezing dream of

> how he kissed me under the Moorish wall and I thought well as
> well him as another and then I asked him with my eyes to ask
> again and he asked me would I yes to say yes my mountain
> flower ... and yes I said yes I will Yes.

And so it is that the creative Yes is exactly the gift I return to my fellow symposiasts.   Noone should be singled out except Peter Gould, who handled everything from the writing of the Rockefeller proposal to the editing of this Pion book.   Yes is indeed the only word I have for capturing the magnificent moments in the Villa, around the terrace, under the tree, on the stonewall at Lago di Como itself.

So what?

Now these fragmentary discourses are bound together, classified, and shelved.   Relations returned to categories.   Perhaps their fate is to gather dust.   But if instead they were to lure just one single student into the forbidden grounds, then this romantic comedy will proceed.

Index[†]

---

† A page number in italic after an author's name indicates a general reference; in upright
type it is a reference to his or her actual work.

Epidemics of uncertainty 73
Epistemology 151, 228
Equilibrium, ecological 114
  market 114
Equivalence relations 237
Eriksson E 74, 170
ERU (Expertgruppen för forskning om
  regional utveckling) 165, 169
Espace de vie 63
Espace vécu 5, 58–59, 63
Essential conditions 47
Essentialism 206
Ethnographic method 66
Ethnoscience 76
Ethological approach 63
Euclidean geometry 116, 125, 127
Euclidean space 59, 116
Evolutionary rhythms 232
Exchange space 198
Exchange value 233–234, 236–238
Existential relationship 58
Existential space 64
Experiential approach 58
Experiential congruence 66
Explanation 80
Extra-genetic mode 256

Factor analysis 98
Familial relations 230
Fast Fourier transform 81
Feedback loops 114
Feibleman H 135
Feuerbach L *190*
Field of constraints 129
  structured 211
Flecker J *72*
Flynn E *247*
Foraminifera 75
Force, attraction 95
  repulsion 95
Ford L 74
Formalization 48
Forrester J 73
Forte A 94
Foucault M 99, 203, 204, 209
Fragmentation 13, 20, 105
  of research 22
Fragmented discourses 153
Framework, *a priori* 39, 49
  meaning 50, 132
  referential 61
  theoretical 47, 54
Frankfurt school 29, 194

Fraser J 257, 258
Frequented space (see *espace de vie*) 63
Freud S *83, 170, 247, 264*
Freudian categories 233
Friction of distance 27
Friedman J 196
Friedmann J 152
Frisby D *45, 205*
Fulkerson D 74
Function 80–81
Functionalism 63, 200
  residual 212
Functionalization 48
Functor 92
Fuzzy set theory 80

Gadamer G 48, 53
Gale S 3
Gallais J 61, 62
Game theory 73
Gaspar J 75, 87
Gatrell A 75, 98
Geertz C 179–180, *226*
Gender research 76, 100
Generative structures 208
Genetic structuralism 200
Geographic theory 106
Geographical materialism 189
Géographie humaine 188, 214
Geography 1, 11, 24–26, 44, 73–74, 106,
  145, 252
  applied 28
  critical 135
  human 157
  qualitative 134–135
  quantitative 134
  phenomenological 42, 65
  radical 154
  regional 32
Geometric determinism 188
Geometric programming 73
Georgi A 42
German geography 25–31
Giddens A 158, 159, *162*, 163, 167, 169,
  171, 173, 175, 176, 178, 180, 194, 198,
  199, 201, *202*, 204–213, *214–215*
Global connectivity 93
Glucksmann M *200*
Goal programming 73
Goddard J 195
Goldblatt R 92
Golledge R *4*, 17, 133
Gorenflo L 98